EVO

TEACHERS GUIDE:

TEN

QUESTIONS
EVERYONE SHOULD ASK ABOUT
EVOLUTION

EVO
TEACHERS GUIDE:
TEN
QUESTIONS
EVERYONE SHOULD ASK ABOUT
EVOLUTION

Rodger W. Bybee and John Feldman

National Science Teachers Association

Arlington, Virginia

NSTApress®
National Science Teachers Association

Claire Reinburg, Director
Jennifer Horak, Managing Editor
Andrew Cooke, Senior Editor
Wendy Rubin, Associate Editor
Agnes Bannigan, Associate Editor
Amy America, Book Acquisitions
    Coordinator

ART AND DESIGN
Will Thomas Jr., Director and
    Frank Carter, cover design
Will Thomas Jr., interior design
Photos courtesy of Hummingbird
    Films LLC, copyright 2010.

Page 37 photo credit:
    Smithsonian Institution

PRINTING AND PRODUCTION
Catherine Lorrain, Director

NATIONAL SCIENCE TEACHERS ASSOCIATION
Francis Q. Eberle, PhD, Executive Director
David Beacom, Publisher
1840 Wilson Blvd., Arlington, VA 22201
www.nsta.org/store
For customer service inquiries, please call 800-277-5300.

FSC
www.fsc.org
MIX
Paper from
responsible sources
FSC® C011935

PERMISSIONS

Library of Congress Cataloging-in-Publication Data
Bybee, Rodger W.
  EVO teachers guide : ten questions everyone should ask about evolution / by Rodger W. Bybee and John Feldman.
      p. cm.
  Includes bibliographical references and index.
  ISBN 978-1-936137-34-3
  1. Evolution (Biology)--Study and teaching. 2. Evolution (Biology)--Miscellanea. I. Feldman, John, 1954- II. Title.
  QH362.B93 2012
  576.8076--dc23
                                2011047201

CONTENTS

PREFACE

This teachers guide for the DVD *EVO: Ten Questions Everyone Should Ask About Evolution* is for science teachers with the dedication, interest, and courage to introduce a unique approach to teaching biological evolution. During the development of *EVO*, we recognized the necessity for a teachers guide that would provide science teachers with insights and support as they implemented an integrated approach to instruction.

This teachers guide provides (1) an introduction to the scientists who are interviewed in *EVO*, (2) background on the 10 questions that form the structure of *EVO*, (3) different ways *EVO* can be implemented, (4) examples of teaching that incorporate the DVD and interviews from *EVO*, and, finally, (5) references and resources for those teachers with the interest and motivation to extend the study of biological evolution.

This project represents collaboration between a filmmaker and a science educator, both interested in biological evolution. The following are brief personal statements from John Feldman, the filmmaker, and then Rodger Bybee, the science educator (curriculum developer).

FROM THE FILMMAKER

When I traveled to the Galápagos Islands in 2005 to lead a workshop making a film about the World Summit on Evolution, little did I know that I would spend the next five years studying evolution and the ways that film can assist in teaching science. My partners, mentors, and the scientists we interviewed at the Summit, as well as other generous and experienced science teachers, educators, students, and members of the scientific community have been incredibly insightful and expressive throughout this journey.

I take the charge of educating through film seriously. This film is based on respect for science and the way that science matters in providing us with a way to respond intelligently, effectively, and ethically to our responsibilities as citizens of the Earth.

Every day's breaking news brings more evidence that we cannot afford to be ignorant about evolution. The study of evolution encourages a deep awareness of the interdependence of all organisms and an appreciation for environmental dynamics through time. It deepens our understanding of the environment and ecology and encourages an unsentimental respect for the natural world.

In making *EVO* I have returned to my first love: nature filmmaking. I made my first film, *A Sense of Existence* (1967), when I was 13; my goal then, as it is now, was to share my fascination and reverence for the natural world.

It is my hope that people will find ways to use *EVO* as a tool for learning about evolution and for further investigations into life in their own backyards.

John Feldman

FROM THE CURRICULUM DEVELOPER

My work at the Biological Sciences Curriculum Study (BSCS), a nonprofit organization that specializes in curriculum development, professional development, and research, included a continuing recognition and inclusion of biological evolution in our science programs. So, the need for and sensitivity to issues associated with teaching biological evolution were not new when John Feldman introduced *EVO*.

I first met John at a National Science Teachers Association (NSTA) meeting. It was in Anaheim, California, in April 2006. He told me about interviews he had of evolutionary biologists and asked how he might produce the film for science teachers.

In my years of work at BSCS, I was approached numerous times with such requests. I politely listened, accepted a DVD that contained several interviews, and told John I would review the interviews. To be truthful, I did not think the discussion would extend beyond this. I was wrong.

About a month later I noticed the DVD on my desk and decided, out of courtesy to John, to spend a few minutes reviewing the material. The quality of the filming, the location—Galápagos Islands—and the biologists left an immediate positive impression. There was no question about the need to do something that made the interviews available to science teachers.

After several unsuccessful attempts to secure funding for a project to finish the videos and prepare a teachers guide, we decided to develop the program on our own.

At BSCS, I learned about the design and development of instructional materials. Obviously, these skills became useful in this project. Work at BSCS also instilled a deep understanding of biological evolution and a profound appreciation for its importance in science education.

Science teachers have a variety of strategies and materials they use to enhance student learning. While use of video is not new or unique, it does present an option that science teachers will find exciting. The opportunity for students to see and hear renowned scientists discussing evolution, while on the Galápagos Islands, is something that will engage the students and enhance their understanding of one of the most influential theories in science.

Rodger Bybee

FROM THE AUTHORS

Like any project, we owe a considerable debt to individuals who provided their assistance. Here we extend our gratitude to Byllee Simon for her assistance and support through the entire project; to Susan Davies, a producer of *EVO*, for her dedication; and to Hugo Burgos and Carlos Montufar at Universidad San Francisco de Quito, which hosted the 2005 World Summit on Evolution.

We also acknowledge individuals who reviewed this guide: Jim Short, director, Gottesman Center for Science Teaching and Learning, Education Department at the American Museum of Natural History; Mark Terry, chair of the science department, Northwest School; and Steve Olson, award-winning author and consultant writer for the National Academy of Sciences and National Research Council. Their insights and recommendations were appreciated and substantially improved the guide.

Finally, we thank Claire Reinberg, Agnes Bannigan, and NSTA for their encouragement and contributions to the guide.

Rodger Bybee
John Feldman

INTRODUCTION

WHY TEACH BIOLOGICAL EVOLUTION?

Questioning the purpose of teaching biological evolution has certainly been paramount for many science teachers. This question holds a significant place in curricular considerations because it encompasses a continuing controversy perpetuated by some religious groups. Unfortunately, teaching biological evolution has become politicized when in fact it should be seen as a nonpartisan, scientifically supported theory. Others have advanced several compelling arguments for this position. For example, science teachers may be interested in Eugenie Scott's *Evolution vs. Creationism* (2004), Brian and Sandra Alters's *Defending Evolution in the Classroom* (2001), the National Academy of Sciences' *Science, Evolution, and Creationism* (2008), James Skehan and Craig Nelson's *The Creation Controversy and the Science Classroom* (2000), Massimo Pigliucci's *Denying Evolution: Creationism, Scientism, and the Nature of Science* (2002), and a compendium of articles for science teachers in *Evolution in Perspective* (Bybee 2004).

Biological evolution is one example of science's intellectual contributions to culture. Indeed, the theory of evolution is among the most significant scientific contributions of the 19th and 20th centuries (Mayr 2000). So, this is one among several reasons to teach biological evolution.

A second reason resides in students' questions about the natural world. After years of observations students may ask, "Why are there so many different kinds of plants and animals?" They also may ask, "How can the similarities of organisms be explained?" Teaching evolution provides scientific answers to these and many other questions that individuals ask about the natural world.

We suggest a third reason. Many life situations that students will encounter as adults are related to basic concepts of biological evolution—for example, taking the full duration of antibiotics for infections. But the importance of understanding basic concepts and processes of biological evolution goes beyond antibiotic resistance. Applications concerning the problem of invasive species and evolving viruses serve as two other examples. Science teachers can find further background and activities in *Evolutionary Science and Society: Educating a New Generation* (Cracraft and Bybee 2005) and *Evolutionary Science and Society: Activities for the Classroom* (Bybee 2006).

Finally, there is a practical reason to teach biological evolution. Biological Evolution: Unity and Diversity will be in the next generation of science standards. The National Research Council's *A Framework for K–12 Science Education: Practices, Crosscutting Concepts, and Core Ideas* (NRC 2011) will serve as the basis for the new standards, and biological evolution is included as a core idea in the life sciences. The core idea, component ideas, and guiding questions in this new framework are the following:

1. **Evidence of Common Ancestry and Diversity:** What evidence shows that different species are related?

2. **Natural Selection:** How does genetic variation among organisms affect survival and reproduction?

3. **Adaptation:** How does the environment influence populations of organisms over multiple generations?

4. **Biodiversity and Humans:** What is biodiversity, how do humans affect it, and how does it affect humans?

Science teachers have many ways of introducing and teaching biological evolution. Options include entire courses, units of study, chapters, and individual activities. *EVO* clips and *EVO Teachers Guide* activities not only complement all of the science teachers' options, but also support science teachers in their efforts to help students meet the standards in scientifically accurate and educationally sound ways.

In addition, seldom do students have the opportunity to hear scientists talk about their work. This may be especially true if their work has to do with one of the greatest intellectual and scientific achievements in human history: the theory of biological evolution. Short descriptions and explanations by scientists, in their own words, will engage students' interest and complement other strategies. Explanations in *EVO* answer basics such as What is evolution? and Who was Charles Darwin? *EVO* also includes discussions such as What is the controversy? and Why should anyone care about evolution?

This teachers guide and the interviews of scientists on the *EVO* DVD provide a portal into one of the greatest contributions of science to society. The opportunities to enhance students' understanding of biological evolution and the processes of science serve as acknowledgement of the courage science teachers demonstrate when they answer the question, Why teach biological evolution?

WHAT IS *EVO*?

EVO is a unique DVD tool for science teachers. The DVD is organized around 10 fundamental questions about biological evolution, and some of the world's best known biologists provide answers to the 10 questions. Interviews of the biologists were gathered on the Galápagos Islands at the 2005 World Summit on Evolution, organized and hosted by the Universidad San Francisco de Quito (USFQ).

EVO is more than interviews, however. The DVD uses footage from the natural world to provide examples of the ideas and processes described by the biologists. Combined with classroom experiences, *EVO* will help students understand some of the most profound and philosophical ideas of science and develop an appreciation of the natural world.

WHO ARE THE SCIENTISTS IN *EVO*?

The World Summit on Evolution was held June 9–12, 2005, in Ecuador's Galápagos archipelago, the islands that helped spark Darwin's revolutionary ideas, which changed how we view life on Earth. The USFQ hosted this conference to celebrate the opening of its Galápagos Academic Institute for the Arts and Sciences (GAIAS). The summit brought together some of the world's preeminent evolutionary biologists and thinkers to discuss and debate current issues in evolutionary biology. The conference consisted of short presentations followed by questions, and then an open discussion between the speakers and participants.

John Feldman was invited to lead a workshop on filmmaking and to document this event. Over the course of the next five years, Mr. Feldman made *EVO*. The scientists who were interviewed for the film all attended the summit, although some of the interviews were shot after the summit. Wildlife footage for the film was shot in the Galápagos and, mostly, in the state of New York.

Scientists

Professor Leticia Aviles (University of British Columbia, Canada) focuses her research on the evolution of sociality, the evolution of sex ratios in subdivided populations, and the evolution of life history traits and local population dynamics. She has published in a variety of journals including *Evolution*, *Ecology*, and *American Naturalist*.

Professor William H. Calvin (University of Washington, United States) is a theoretical neurobiologist and the author of a dozen books, mostly for general readers, about the brain and evolution, including *A Brief History of the Mind: From Apes to Intellect and Beyond* (Oxford 2004). His book with Derek Bickerton, *Lingua ex Machina: Reconciling Darwin and Chomsky With the Human Brain* (MIT 2000), is about the evolution of structured language.

Professor Daniel C. Dennett (Tufts University, United States) is a noted philosopher whose research focuses on philosophy of mind, philosophy of science, and philosophy of biology. He is the author of more than a dozen books, including *Darwin's Dangerous Idea* (Simon & Schuster 1995). He has received two Guggenheim Fellowships, a Fulbright Fellowship, and a Fellowship at the Center for Advanced Studies in Behavioral Science. He was elected to the American Academy of Arts and Sciences in 1987.

Dr. Kevin de Queiroz (Smithsonian Institution, United States) is a research zoologist and curator at the National Museum of Natural History. His current research centers on the phylogenetic relationships of various groups of lizards. He is also interested in phylogenetic nomenclature and is the co-originator of the PhyloCode, which has been proposed to replace Linnaean taxonomy.

Dr. Niles Eldredge (American Museum of Natural History, United States) is a paleontologist and curator of the American Museum of Natural History. With

Stephen Jay Gould, he coauthored the theory of punctuated equilibria, a milestone in evolutionary theory. His specialty is the evolution of trilobites. His books include *Darwin: Discovering the Tree of Life* (W. W. Norton 2005), *Why We Do It: Rethinking Sex and the Selfish Gene* (W. W. Norton 2004), *The Triumph of Evolution and the Failure of Creationism* (Holt 2000), and *Reinventing Darwin: The Great Debate at the High Table of Evolutionary Theory* (Wiley 1995).

Professor Douglas Futuyma (Stony Brook–State University of New York, United States) is author of the popular textbooks: *Evolutionary Biology* (Sinauer Associates, 3 editions), *Evolution* (Sinauer Associates 2005, 2009) and *Science on Trial: The Case for Evolution* (Sinauer Associates 1982, 1995). His research concerns the speciation of insects and the evolution of interactions between herbivorous insects and their host plants. He is the editor of *Annual Review of Ecology, Evolution and Systematics* and has been the editor of *Evolution*. He has been president of the Society for the Study of Evolution and the American Society of Naturalists, from which he received the Sewall Wright Award. He was made a member of the National Academy of Sciences in 2006.

Professor Pierre-Henri Gouyon (Université Paris-Sud, France) is recognized for his work on evolution and genetics. He is coauthor of *Gene Avatars: the Neo-Darwinian Theory of Evolution* (Belin 1996; English translation, Springer 2002). He is the subject of a film by Nicholas Ribowski titled *Pierre-Henri Gouyon: Génétique et Evolution* (2005). He is a professor at the Muséum National d'Histoire Naturelle de Paris and at Sciences Po, Paris (École libre des sciences politiques), and he is managing editor of the European Society of Biological Evolution's *Journal of Evolutionary Biology*.

Professor Peter Grant (Princeton University, United States) and **Professor Rosemary Grant** (Princeton University) are widely known for their remarkable long-term studies of more than 35 years, which demonstrate evolution in action on Darwin's finches on the Galápagos Islands. In 2005, they received the Balzan Prize for Population Biology that praises their "seminal influence in the field of population biology, evolution, and ecology." In 2008, Peter and Rosemary Grant were among the 13 recipients of the Darwin–Wallace Medal, which is presented every 50 years by the Linnean Society of London. In 2009, they were recipients of the annual Kyoto Prize in basic sciences, an international award honoring significant contributions to the scientific cultural and spiritual betterment of mankind.

Professor Laura Katz (Smith College, United States) researches the eukaryotic tree of life, the phylogeography of coastal ciliates, and genome evolution in microbial eukaryotes. She is the editor of the book *Genomics and Evolution of Microbial Eukaryotes* (Oxford 2006) and was associate editor for the journal *Molecular Biology and Evolution* (2003–2008). She was on the scientific advisory board for the National Evolutionary Synthesis Center (NESCent) when it was newly formed (2005–2007). In addition to teaching at Smith, she is on the graduate faculty at the University of Massachusetts–Amherst.

Professor **Antonio Lazcano** (Universidad Autónoma de México, Mexico) is a leading scholar in the study of the origin and early evolution of life and is the author of *El origen de la vida* (*The Origin of Life* 1984). He is the first Latin American scientist to have served as president of the International Society for the Study of the Origin of Life (ISSOL).

Professor **Lynn Margulis** (University of Massachusetts–Amherst, United States) is renowned for her original contributions to the study of microbial evolution and cell biology. She is best known for her theory on the origin of eukaryotic organelles and her contributions to the endosymbiotic theory. Dr. Margulis contributed to James E. Lovelock's Gaia concept. Yale ecologist G. E. Hutchinson credited her with creating a "quiet revolution in microbiological thought." She received the National Medal of Science from President Clinton and Germany's Alexander von Humboldt Prize. In 2008, she was one of only 13 recipients of the Darwin–Wallace Medal, which is granted every 50 years by the Linnean Society of London. She coauthored a number of books with her son Dorion Sagan, including *Symbiotic Planet: A New Look at Evolution* (Basic Books 1998) and *Acquiring Genomes: A Theory of the Origins of Species* (Basic Books 2002). Lynn Margulis died on November 22, 2011. She was an active supporter of *EVO* and contributed much to its making. She was a courageous scientist, and we will miss her.

Professor **Geoff McFadden** (University of Melbourne, Australia) studies protists (eukaryotic microorganisms) and endosymbiosis. He has identified the relict chloroplast in malaria parasites and is developing herbicides as antimalarial drugs. He has published in journals such as *Nature, Science, EMBO J,* and *PNAS.* Professor McFadden has been awarded, among others, the Australian Academy of Science's Frederick White Prize, two Howard Hughes Medical Institute Scholar's awards, and the Royal Society of Victoria Medal. He is a member of the Australian Academy of Sciences.

Professor **Richard Michod** (University of Arizona, United States) researches the evolution of interactions within populations, particularly cooperative interactions and conflict. He has specifically examined the evolution of sex, origin of individuality, origin of life, and the evolution of social behavior. He has written *Dawinian Dynamics* (Princeton 1999), *Eros and Evolution* (Perseus 1996), and *A Natural Philosophy of Sex* (Addison Wesley 1995). He has also edited books on the evolution of sex and human values.

Professor **Joan Roughgarden** (Stanford University, United States) currently focuses her research on linking ecology with economic theory. Her books include *The Genial Gene: Deconstructing Darwinian Selfishness* (University of California Press 2009), *Evolution and Christian Faith: Reflections of an Evolutionary Biologist* (Island Press 2006), and *Evolution's Rainbow: Diversity, Gender, and Sexuality in Nature and People* (University of California Press 2004, 2009).

Professor **Frank Sulloway** (University of California–Berkeley, United States) has written about the nature of scientific creativity and published extensively on the life and theories of Charles Darwin. He is a visiting scholar in the Institute of Personality and Social Research at the University of California–Berkeley. His book, *Freud, Biologist of the Mind: Beyond the Psychoanalytic Legend* (Harvard 1979) received the Pfizer Award of the History of Science Society. Dr. Sulloway is a former MacArthur Fellow, a Fellow of the American Association for the Advancement of Science, and a recipient of the Golden Plate Award of the American Academy of Achievement.

WHAT IS THE STRUCTURE OF *EVO*?

The *EVO* DVD has 10 short videos, each based on a question. The following paragraphs summarize the 10 questions and segments and include a running time for each.

Question 1: What is evolution? (11 mins. 44 secs.)

EVO begins with a brief introduction to the World Summit on Evolution, which was the genesis for this film, and then presents the questions, What is evolution? and How long does evolution take? *EVO*'s signature arm graphic helps viewers visualize the enormous amount of time involved in the evolution of life.

Question 2: Who was Charles Darwin? (11 mins. 15 secs.)

How does Charles Darwin fit into the picture? In this brief portrait of Darwin, Frank Sulloway helps trace both the physical and mental journey that led Darwin to the idea that all organisms have evolved from a single common ancestor through natural selection.

Question 3: What is natural selection? (15 mins.11 secs.)

A joke from philosopher Daniel Dennett begins this investigation into examples of natural selection—some in the Amazon forest, some on the Galápagos Islands, but most in the filmmaker's own backyard in the mountains and forests of New York. *EVO*-style animation is used to illustrate the process, while Douglas Futuyma and some squawking birds named cocks-of-the-rock help unravel the puzzle of sexual selection.

Question 4: How do species come about? (11 mins. 1 sec.)

What is a species? Douglas Futuyma and Kevin de Queiroz tackle the "origin of species" question, and then *EVO*-style animation shows us how scientists use phylogenetic trees to visualize species splitting over time. Laura Katz clears up two common misconceptions by making it plain why humans are not descendants from chimps and why humans are not the "pinnacle of evolution."

Question 5: Where do variations come from? (9 mins. 40 secs.)

If the differences—the variations—between two individuals of the same species are the raw material for natural selection, where do these variations come from? Douglas Futuyma and William Calvin give brief introductions into genetics, DNA mutation, and sexual recombination. Then Lynn Margulis introduces the concept of symbiogenesis.

Question 6: What role does cooperation play? (9 mins. 43 secs.)

"If we form alliances with others, we can do more than going it alone," says Richard Michod as he explains that cooperation is just as important a driver of evolutionary change as competition, which usually gets all the credit. *EVO* looks at examples of cooperation ranging from human cells to ants, and hyenas to human society. Lynn Margulis introduces the Gaia idea and helps viewers see that we are all connected and part of a complex and robust Earth system.

Question 7: What is a brief history of life? (8 mins. 45 secs.)

There is evidence that the Earth formed 4.5 billion years ago and that life began approximately 3.6 billion years ago. In exactly 4.5 minutes this *EVO*-style animated film takes us along a timeline from the beginning of the Earth to the present. Niles Eldredge then reminds viewers what a young species we are and describes "punctuated equilibrium," a theory first proposed by him and Stephen J. Gould in 1972.

Question 8: What is the controversy? (9 mins. 59 secs.)

A public controversy continues over whether creationism can be taught in public school science classes along with or in place of biological evolution. Scientist Michod and philosopher Dennett look at the philosophical roots of the controversy and trace the history of the controversy with the help of Hollywood and Susan Epperson, the biology teacher who made a Supreme Court challenge in support of teaching evolution in the schools.

Question 9: Is evolution random? (8 mins. 4 secs.)

If genetic mutations occur at random, isn't evolution random? First, Douglas Futuyma defines the word *random* and clarifies the view that natural selection—the driving force of evolution—is not at all random. Daniel Dennett then describes natural selection as a sorting algorithm that is, like a computer, totally mindless. *EVO*-style animated snails, along with a careless cow, return to illustrate Futuyma's explanation of genetic drift.

Question 10: Why should anyone care about evolution? (11 mins. 36 secs.)

Evolution just happens, so why is it relevant to me? In answer to this final question, the scientists are quick to point out that we must understand evolution if we care about fighting disease and if we want to learn to live in a sustainable balance with our environment instead of degrading it. After looking at the biotechnology industry and introducing the idea of cultural evolution, John Feldman concludes *EVO* by reflecting on the importance of cooperation within the history of life and the choice each of us has to respect and care for the ecosystem of which we are a part.

HOW CAN SCIENCE TEACHERS USE *EVO*?

EVO can be used in many different ways. The only real limit is the science teacher's professional creativity. Several uses seem obvious:

1. Science teachers could use *EVO* as a unifying theme during a yearlong life science or integrated science class. Science teachers can make decisions about the best places to insert the interviews and activities within the curriculum. The *EVO* DVD and the 10 questions do not have to be used as a total program.

2. *EVO* easily complements a unit on biological evolution. Biology and life science textbooks typically include a unit on evolution. Depending on the science teacher's discretion, different videos (i.e., questions) from *EVO* could be introduced as supplements to the curriculum. The videos and their respective lessons in this teachers guide can be used in any order. For example, an application of the BSCS 5E Instructional Model (see "Ten Lessons Using *EVO*" on p. xxi) could be applied to the 10 questions and the videos used as the basis for an instructional unit. Based on the 10 questions and activities developed for this guide, the following reorganization illustrates an instructional unit:

   - Engage
     Question 1: What is evolution?
     Question 8: What is the controversy?

   - Explore
     Question 2: Who was Charles Darwin?
     Question 3: What is natural selection?
     Question 7: What is a brief history of life?

   - Explain
     Question 4: How do species come about?
     Question 5: Where do variations come from?

   - Elaborate
     Question 6: What role does cooperation play?
     Question 9: Is evolution random?

   - Evaluate
     Question 10: Why should anyone care about evolution?

3. Science teachers could use the 10 videos and complementary activities as a short course. Using the DVD as a central feature and adding investigations, readings, and classroom presentations, the lessons and resources included in this guide could serve as the supplements to create such an introductory course. The lessons included in this guide are only intended as optional supplements for science teachers.

4. Finally, science teachers could use the interviews as the basis for students' independent study. Science teachers would use their judgment about the parameters of individualized or differentiated study, but the interviews and questions certainly provide an excellent place for students to begin study as well as begin a range of research projects identified within the *Look-Again* markers in the *EVO* DVD.

The *Look-Again* markers are an index to *EVO*'s significant thoughts and moments. Clicking on a *Look-Again* menu item takes you to that point in *EVO*. To stop playing and return to the *Look-Again* menu, push Menu on your remote control. While playing *EVO*, you can use the Forward, Next, or Next Chapter button your remote to jump from one marker to the next.

TEN LESSONS USING *EVO*

The 10 lessons for *EVO* have a clear focus on different aspects of evolution. Concepts such as natural selection and genetic variation as well as understanding the scale of time for evolution are the learning outcomes. The lessons have two additional features: (1) They are based on the 5E Model, which will enhance student learning of the biological concepts, and (2) the lessons include opportunities that will help students develop skills important in the 21st century. The following discussion provides a background on the 5E Model and 21st-century workforce skills found in the *EVO Teachers Guide* lessons.

THE BSCS 5E INSTRUCTIONAL MODEL

The *EVO Teachers Guide* lessons incorporate the 5E Model. The instructional model includes five steps: (1) engagement, (2) exploration, (3) explanation, (4) elaboration, and (5) evaluation of science concepts and processes. The following discussion provides details of the five stages.

ENGAGE THE STUDENTS.

This stage of the instructional model initiates the learning task. The activity makes connections between past and present learning experiences and anticipates activities and helps focus students' thinking on the learning outcomes of current activities. Students should become mentally engaged in the concept, process, or skill of the lesson.

Scientific investigations of biological evolution originate with a question that engages a scientist, so too must students engage in the activities of learning. The *EVO Teachers Guide* lessons begin with a strategic question that gets students thinking about the content of the lesson.

STUDENTS EXPLORE THE TOPIC.

This stage of the instructional model provides students with a common base of experiences within which they identify and develop current concepts, processes, and skills. During this stage, students actively explore their environment or manipulate materials.

Once engaged, students need time to explore ideas before concepts begin to make sense. In this exploration stage, teachers provide opportunities for students to try their ideas, ask questions, and look for possible answers to questions. In the *EVO Teachers Guide*, students use inquiry strategies and try to relate their ideas to those of other students and to what biologists already know about evolution.

STUDENTS, TEACHERS, AND SCIENTISTS PROVIDE <u>EXPLANATIONS</u>.

This stage of the instructional model focuses students' attention on a particular aspect of their engagement and exploration experiences and provides opportunities for them to develop explanations and hypotheses. This phase also provides opportunities for teachers to introduce a formal scientific label or definition for a concept, process, or skill.

In the *EVO Teachers Guide* lessons' third step, students propose answers and develop hypotheses. Also in this step, *EVO* has an especially strong place for introducing the scientists and their explanations of the various aspects of biological evolution. This also is the step when teachers should make the major concepts explicit and clear to the students.

STUDENTS <u>ELABORATE</u> THEIR UNDERSTANDINGS AND SKILLS.

This phase of the instructional model challenges and extends students' conceptual understanding and allows further opportunity for students to test hypotheses and practice desired skills. Through new experiences, the students develop a deeper and broader understanding, acquire more information, and develop and refine skills.

Science teachers understand that informing students about a concept does not necessarily result in their immediate comprehension and understanding of the idea. So the *EVO Teachers Guide* lessons provide a step referred to as elaboration in which students have opportunities to apply their ideas in new and slightly different situations.

<u>EVALUATE</u> STUDENTS' UNDERSTANDING AND SKILLS.

This stage of the instructional model encourages students to assess their understanding and abilities and provides opportunities for teachers to evaluate student progress toward achieving the learning outcomes.

Teachers need to assess how well students understand the concepts, or how successful they are at applying the desired skills. These are the questions to be answered during the evaluation stage. In the *EVO Teachers Guide* lessons, we have provided a single rubric for science teachers to use in the evaluation stage.

This use of the 5E Model usually centers on a daily lesson as the parameters for the instructional sequence. In other cases, the instructional sequence is longer, perhaps several days or a series of lessons that amount to a unit of study. Although a longer instructional sequence provides greater opportunities for students to learn, we wanted to focus on the *EVO* interviews, without assuming an extended time, and leave additional activities to the science teacher's professional discretion. Each of the questions (i.e., DVD segments) represents a topic of study that is appropriate for individual or group work.

21st-CENTURY WORKFORCE SKILLS

These lessons for the *EVO* DVD provide opportunities to develop different skills and abilities that students will need as they prepare for college and careers. The opportunities to develop the following skills present themselves in any inquiry-oriented classroom where students interact, problem solve, communicate findings, complete homework, and engage in serious thought about scientific explanations about evolution.

ADAPTABILITY

In some *EVO* lessons, students will be required to cope with new approaches, grapple with new explanations, and work with new tools and techniques. In addition, in some cases, students will work in groups. In all of these cases, science teachers have the opportunity to help students develop adaptability.

COMPLEX COMMUNICATIONS/SOCIAL SKILLS

Lessons with varied learning experiences, including investigations, will require students to process and interpret information and data from a variety of sources. Learners would have to select appropriate evidence and use it to communicate a scientific explanation. Some *EVO Teachers Guide* lessons include group work that culminates with the use of information to formulate a conclusion or recommendation.

NONROUTINE PROBLEM SOLVING

Some lessons require learners to apply knowledge to scientific questions and technological problems, identify the scientific components of a contemporary issue, and use reasoning to link evidence to an explanation. Students may be required to think of another explanation, or another way to gather data, and connect those data with their knowledge.

SELF-MANAGEMENT/SELF-DEVELOPMENT

EVO Teachers Guide lessons include opportunities for students to work on activities individually and collaboratively. These activities require learners to acquire new knowledge and develop new skills as they pursue answers to questions or solutions to problems.

SYSTEMS THINKING

Some *EVO Teachers Guide* lessons include the introduction and applications of systems thinking in the context of life science as well as multidisciplinary problems in personal and social perspectives. Learners may be required to realize the limits to systems; to describe components, flow of resources, and changes in systems and subsystems; and to reason about interactions at the interface between living systems.

Table 1 summarizes essential features of the skills and provides examples for school science programs.

As you will see, not all skills are a part of each lesson. Only those skills appropriate to the *EVO Teachers Guide* lesson are presented.

Table 1. Developing 21st-Century Skills in Science Programs

Essential Features of 21st-Century Skills	Examples of Contexts for School Science Programs
Adaptability	
• Cope with changing conditions • Learn new techniques, procedures • Adapt to different personalities and communication styles • Adapt to different working environments	• Work on different investigations and activities • Work on different activities • Work cooperatively in groups • Work on lessons in different situations
Complex Communications/Social Skills	
• Process and interpret verbal and nonverbal information • Select key pieces of complex ideas to communicate • Build shared understanding • Negotiate positive outcomes	• Prepare oral and written reports communicating procedures, evidence, and explanations of investigations and activities • Use evidence gained in investigations as the basis for a scientific explanation • Prepare a scientific argument • Work with group members to prepare a report
Nonroutine Problem Solving	
• Use expert thinking in problem solving • Recognize patterns • Link information • Integrate information • Reflect on adequacy of solutions • Maintain several possible solutions • Propose new strategies • Generate innovative solutions	• Recognize the need for an expert's knowledge • Recognize patterns in data • Connect evidence and information from an investigation with scientific knowledge from the video, teachers, or other sources • Understand constraints in proposed solutions • Propose several possible solutions and strategies to attain the solutions • Propose creative solutions
Self-Management/Self-Development	
• Work remotely (individually) • Work in virtual teams • Develop self-motivation • Develop self-monitoring • Display willingness and ability to acquire new information and skills	• Work individually at home • Work with a virtual group • Complete a full/open activity • Reflect on adequacy of progress, solutions, and explanations • Acquire new information and skills in the process of problem solving and working on an investigation
Systems Thinking	
• Understand the systems concept • Understand how changes in one part of the system affects the system • Adapt a "big picture" perspective • Complete system analysis • Demonstrate judgment and decision making • Apply abstract reasoning about interactions among components of a system	• Describe components of a system based on a system under investigation • Predict changes in an investigation • Analyze a system under investigation • Make decisions about best proposed solutions • Demonstrate understanding about components and functions of a proposed system

LESSON ONE:
What Is Evolution?

OVERVIEW

This lesson engages students in the concepts and processes of biological evolution. It also introduces the *EVO* DVD. The lesson begins by viewing Question 1 of the DVD, which introduces the World Summit on Evolution and asks the question What is evolution? Students present their ideas and answers to the lead question. The lesson then returns to the DVD for several scientists' answers to the lead question. Students work in small groups to develop a one-paragraph answer and questions they have about evolution.

Time: One 50- to 55-minute class period   **DVD: 11 mins. 44 secs.**

LEARNING OUTCOMES

As a result of this lesson, students should be able to

- develop initial knowledge of biological evolution,
- demonstrate some appreciation for scientists' views about biological evolution, and
- develop communication and social skills.

MATERIALS FOR THE LESSON

- Equipment to play the *EVO* DVD for the entire class and computer workstations for students to view the *EVO* DVD during group classwork. Arrange groups of four to five students each.

BACKGROUND FOR THE TEACHER

Biological evolution is one of the most powerful and influential scientific explanations in the history of science. Consider the following brief summary: Past and present life forms have evolved from common ancestors, the lineage of which extends back in history to simple one-celled organisms. Evolution provides a scientific explanation for three observations about life: (1) There exists a remarkable diversity of life on Earth; (2) systemic similarities of anatomy, physiology, and the molecular basis of life presents a paradoxical unity of life; and (3) the sequence of changes in fossils formed during geological history.

The goal of this first lesson in the series should be for students to establish an initial understanding of evolution. In many cases, you will observe that the students' current understanding may best be described as misconception. For example, students have a difficult time with the idea that evolution occurs across generations and in populations, not to individual organisms during the course of their lifetimes. You also may observe some other misconceptions: a belief that environmental conditions are responsible for changing traits, individual organisms develop new traits in order to survive, or individual organisms' use and disuse of organs or abilities will result in new traits in the next generation.

The continued exposure to scientifically accurate concepts through structured activities and explanations will contribute to students' learning of biological evolution. The structure of these *EVO Teachers Guide* lessons and the DVD interviews will complement other activities and contribute to student learning.

PREPARING FOR THE LESSON

- Review the *EVO* DVD. Go to the menu's Select A Question and review Question 1, What is evolution?
- Set up equipment (e.g., a computer or DVD player) for class presentation.
- Arrange for student computers (e.g., one for each group of four to five students).
- Set up student computers.

LEADING THE LESSON

ENGAGE the Students. (4 mins. for DVD segment)

1. In a brief statement, tell the students that they are going to see a DVD segment that introduces biological evolution.
2. Show an initial segment of *EVO*'s Question 1, What is evolution?
3. Stop the DVD after four minutes, as the image of scientists at the World Summit walking out on rocks appears.

Students EXPLORE the Idea of Evolution. (10 mins.)

1. Ask students to write their answer to the question, What is evolution?
2. Have several students share their answers to the question. You should accept their answers.
3. Tell the students that we will now see how scientists answer the question.

EXPLAIN the Idea of Evolution. (8 mins. for DVD segment; 5 mins. for questions)

1. Return to the DVD and start where you stopped. This should begin with the question, What is evolution?
2. Continue playing the DVD until the end of the talk section.
3. Scientists will provide a variety of answers to the question, What is evolution?
   - Common descent
   - Organisms are descended from a common ancestor.
   - All organisms are related to a common ancestor.
   - Common descent of living forms
   - With common descent things change.
   - Change over time
   - Evolution is the change in various characteristics of organisms.
4. The DVD introduces the scale of time and the idea that the environment, the atmosphere, and populations of organisms have all changed concurrently through geological time.
5. Continue the lesson by clarifying students' questions.

Students ELABORATE the Idea of Evolution. (25 mins.)

1. Students work in groups of four or five. Their task is to (1) develop a short statement answering the question, What is evolution? and (2) identify three to four questions they have about biological evolution.

2. The students may want to review sections of the DVD. Direct them to use the *Look-Again* markers. For this segment of the DVD, the *Look-Again* markers will direct viewers to the following clips:
   - tent caterpillars, diversity
   - World Summit on Evolution
   - what is evolution?
   - indian pipes, evolution defined
   - arm timeline, how long … ?
   - Grand Canyon

3. Have the student groups present their answers and questions about evolution.

EVALUATE Students' Learning.

Use the following criteria for evaluating students' learning:

Exceeds expectations: All outcomes are met, work goes beyond expectations and demonstrates exceptional understanding.

Meets expectations: Work indicates that student understands major issues and concepts.

Below expectations: Student work does not meet the criteria. It may be incomplete or may not address key issues or concepts of the learning outcomes.

Learning Outcomes	Evaluation
Develop initial knowledge about biological evolution	
Demonstrate some appreciation of scientists' perspectives	
Develop communication and social skills	
Overall evaluation for this lesson	

LESSON TWO:
Who Was Charles Darwin?

OVERVIEW

This lesson begins by asking students the questions, Who was Charles Darwin? and What do you know about him? After collecting students' current knowledge about Darwin, the *EVO* DVD provides background on the explorations and experiences that resulted in Darwin's formulation of the idea that all organisms evolved from a single ancestor through natural selection. The lesson concludes with students extending their knowledge of evolution by examining actual statements by Darwin and other historical figures.

Time: Two 50- to 55-minute class periods   **DVD: 11 mins. 15 secs.**

LEARNING OUTCOMES

As a result of this lesson, students should be able to

- develop knowledge about Charles Darwin and his contribution to science,
- develop an understanding of the nature of science as seen in the formulation of the theory of evolution, and
- develop skills of self-management and communication.

MATERIALS FOR THE LESSON

- Equipment to play the *EVO* DVD for the entire class and computer workstations for students to view the *EVO* DVD during group classwork. Arrange groups of four to five students each.
- Handouts of excerpts by Jean Lamarck, Alfred Russel Wallace, and Charles Darwin (see Appendix 1.1, 1.2, and 1.3 on pp. 65–68)

BACKGROUND FOR THE TEACHER

Figure 1. A Summary of Charles Darwin's Life

Charles Darwin
February 12, 1809–April 19, 1882

1809	Born to Robert Darwin and Susannah Wedgewood, Shrewsbury, Shropshire, England.
1825	Begins studies to become physician, Edinburgh University
1828	Continues studies to become a minister, Christ's College, Cambridge University
1830	Graduates with a bachelor of arts degree, Cambridge University
1831	Graduates with a master of arts degree, Cambridge University
1836	Departs England on HMS *Beagle* and serves as a naturalist collecting specimens and a gentleman companion to Captain Robert FitzRoy (December 27)
1836	Returns to England after voyage of the *Beagle*
1837	Formulates his tree of life theory (July)
1838	Formulates his theory of natural selection (September)
1839	Marries Emma Wedgewood, his cousin (January 29)
1839	Publishes *The Voyage of the Beagle* (June)
1842	Moves from London to Down House, Kent England
1858	Presents (in absentia) with naturalist Alfred Russel Wallace the first public statements on the theory of evolution at the Linnaean Society of London (July 1)
1859	Publishes *On the Origin of Species by Means of Natural Selection, or the Preservation of Favoured Races in the Struggle for Life.* The title later was shortened to *The Origin of Species by Means of Natural Selection* (1872).
1882	Dies at Down House, age 73. Buried in Westminster Abbey, London

Who was Charles Darwin, and what did he do? Figure 1 shows a summary of Darwin's milestones throughout his life. Although the idea of evolution had long been discussed (e.g., by some Greeks and Darwin's grandfather, Erasmus), Darwin is credited with formulating the theory of biological evolution, one of the most significant ideas in the history of science. Darwin asked three fundamental questions about life on Earth: (1) How can one explain the diversity of species? (2) How can common features among species be explained? (3) How do species become so suited to their surroundings? In brief, Darwin answered questions about the diversity, unity, and adaptation of species.

PREPARING FOR THE LESSON

- Review the *EVO* DVD. Go to the menu's Select A Question and review Question 2, Who was Charles Darwin?
- Set up equipment (e.g., a computer or DVD player) for class presentation.
- Prepare copies of the three excerpts from Lamarck, Wallace, and Darwin (i.e., one copy per student). See Appendix 1.2, 1.2, and 1.3 on pages 65–68.

LEADING THE LESSON

ENGAGE the Students. (15 mins.)

1. Begin by asking the students if they have ever heard of Charles Darwin.
2. Follow their responses by asking what they know about Darwin. Ask questions such as Where did he live? and What did he do?

EXPLORE the Life of Charles Darwin. (30 mins.)

1. Briefly tell the students the story that answers the questions, Who was Charles Darwin? and What did he do?
2. Use the information from Background for the Teacher for the story.
3. Tell the story as a way to set up the DVD segment, which emphasizes Darwin's voyage on the *Beagle*, exploration of the Galápagos Islands, and formulation of his theory of biological evolution.

EXPLAIN Who Darwin Was and His Formulation of the Theory of Evolution by Natural Selection. (11 mins. 15 secs. for DVD segment)

1. Show Question 2 on the *EVO* DVD, Who was Charles Darwin?
2. After the DVD segment, continue the lesson by answering student questions.
3. You may want to use the *Look-Again* markers during the discussion. The *Look-Again* markers will direct viewers to the following clips:
   - marine iguanas

- mockingbird speciation animation
- Darwin's big breakthrough

ELABORATE Charles Darwin's Idea of Evolution. (50–55 mins., or one class period)

1. This phase of the lesson clarifies Darwin's scientific explanation for evolution.

2. Students read and discuss three brief explanations for the diversity of species by Jean Lamarck, Alfred Russel Wallace, and Charles Darwin (see Appendix 1.1, 1.2, and 1.3 on pp. 65–68).

3. Distribute the three excerpts. Each student gets one copy of each excerpt.

4. Assign students the three readings as homework.

5. On the second day of the lesson, students work in groups of four and discuss the three readings. The intention of this portion of the lesson is to build on the prior day and further develop students' understanding of Darwin's theory by providing historical context for its development. In addition, you should use these readings and discussions to emphasize the nature of scientific explanations.

6. Have students begin by discussing Jean Lamarck's explanations for changes in organisms. You should be aware that some students may hold the Lamarckian view as their idea of evolution. Questions you might pose include the following:
   - What role does the environment play in Lamarck's explanation?
   - Was Lamarck's explanation scientific? Why or why not?
   - Can you propose another explanation for Lamarck's observation about the disuse and use of organs?
   - Can you think of any examples in which acquired characteristics are passed down to offspring?

7. Next, direct students to discuss the excerpt by Alfred Russel Wallace. The following questions should guide student discussions:
   - Wallace proposed that the life of wild animals is a struggle for existence. Is Wallace's view scientific? Why or why not?
   - Wallace claims that "useful variations will tend to increase, unuseful or hurtful variations to diminish." Does he cite any evidence for this? What processes account for these changes?
   - How does Wallace's explanation differ from Lamarck's?

8. Finally, still working in groups, have students refer to the excerpt from *On the Origin of Species* by Charles Darwin. The intention is to synthesize discussions and the *EVO* DVD with Darwin's actual words. You might use the following questions to guide student discussions:
   - How did the voyage and explorations on H.M.S *Beagle* contribute to Darwin's formulation of his theory of evolution?

- On what did he base his explanations?
- What was the relationship of Lamarck's and Wallace's work to Darwin's?
- Was Darwin's explanation scientific?

EVALUATE Students' Learning.

1. As homework, have each student write a brief essay titled, Who was Charles Darwin and What Questions Did He Ask and Answer? Have students observe several organisms in their neighborhood and ask questions that Darwin might ask (e.g., How are these organisms suited to their surroundings?). The essay should include:
   - three quotes from the readings (one from Darwin is required),
   - an indication of the scientific nature of Darwin's ideas, and
   - reference to the *EVO* DVD.

2. Based on your review of each student's essay, you should determine the degree to which he or she has met the learning outcomes.

3. Use the following criteria for evaluating students' learning:

   Exceeds expectations: All outcomes are met, work goes beyond expectations and demonstrates exceptional understanding.

   Meets expectations: Work indicates that student understands major issues and concepts.

   Below expectations: Student work does not meet the criteria. It may be incomplete or may not address key issues or concepts of the learning outcomes.

Learning Outcomes	Evaluation
Develop knowledge of Charles Darwin's contributions to science	
Develop an understanding of the nature of science as seen in the formulation of the theory of evolution	
Develop self-management skills	
Develop communication skills	
Overall evaluation for this lesson	

LESSON THREE:
What Is Natural Selection?

OVERVIEW

This lesson uses an invitation to inquiry strategy to answer the question, What is natural selection? The students are engaged by a problem of a farmer trying to eliminate flies from a barn. Students explore reasons for the return of the flies. After a brief explanation of natural selection, the *EVO* DVD is used to extend and deepen students' understanding of natural selection, a basic idea in understanding evolution. This activity is adapted with permission from *The Biology Teachers Handbook, 4th Edition*.

Time: One 50- to 55-minute class period   **DVD: 15 mins. 11 secs.**

LEARNING OUTCOMES

As a result of this lesson, students should be able to

- develop knowledge about the process of natural selection,
- develop their ability to create explanations based on evidence, and
- develop skills of complex problem solving and communication.

MATERIALS FOR THE LESSON

- Equipment to play the *EVO* DVD for the entire class

BACKGROUND FOR THE TEACHER

There is no better person to present background on natural selection than Darwin himself, using his words from his 1859 publication *On the Origin of Species*. In the opening chapters of *On the Origin of Species*, Darwin discusses domestic breeding, variation in species, and the struggle for existence. Near the beginning of Chapter 3, Darwin introduces natural selection:

> Can it, then, be thought improbable, seeing that variations useful to man have undoubtedly occurred, that other variations useful in some way to each being in the great and complex battle of life, should sometimes occur in the course of thousands of generations? If such do occur, can we doubt (remembering that many more individuals are born than can possibly survive) that individuals having any advantage, however slight, over others, would have the best chance of surviving and of procreating their kind? On the other hand, we may feel sure that any variation in the least degree injurious would be rigidly destroyed. This preservation of favourable variations and the rejection of injurious variations, I call Natural Selection. (pp. 80–81)

In this quotation, Darwin presents the basic idea that individuals within populations of organisms have, as we know today, heritable variations. He also presents the idea that within any population, more organisms are born than current resources and their environments can support. This situation favors those organisms with traits that give them a survival advantage and thus the opportunity to reproduce.

Here is a succinct summary of the key ideas that form the theory of biological evolution from the *National Science Education Standards* (NRC 1996):

> Evolution is the consequence of the interaction of (1) the potential for a species to increase in number, (2) the genetic variation among

offspring due to mutation and recombination of genes, (3) a finite supply of the resources and stable environments required for life, and (4) the ensuing selection of those offspring better able to survive limited resources and changing environments and leave offspring. (p. 185)

Students have difficulty with the fundamental concepts of evolution. For example, some students express misconceptions about natural selection because they do not understand the relationship between variations within a population, change in individual characteristics during the life of the individual, and change in the populations of individual variations over time.

This activity emphasizes natural selection. In particular, it presents students with a significant environmental change—an insecticide—as one example of how natural selection operates in nature.

Students should understand that the ten process of evolution has two steps, referred to as *genetic variation* and *natural selection*. The first step is the development of genetic variation through changes such as genetic recombination, gene flow, and mutations. The second step, and the point of this activity, is natural selection.

There are a variety of natural selection simulations and games that can help students understand this concept, such as Understanding Evolution at *http://evolution. berkeley.edu*.

PREPARING FOR THE LESSON

- Review the *EVO* DVD. Go to the menu's Select A Question, and review Question 3, What is natural selection?

- Set up equipment (e.g., a computer or DVD player) for class presentation.

LEADING THE LESSON

ENGAGE the Students. (5 mins.)

1. Have the students work in pairs.

2. Begin by engaging the students with the problem and the basic information they will need to begin solving the problem.

3. To the students: *A farmer sprayed his barn and cattle with a solution of Insecticide A. The insecticide killed nearly all the flies. Sometime later, however, the number of flies was again large. For a second time, the farmer sprayed with Insecticide A. The result was similar to that of the first spraying: Most, but not all, of the flies were killed. Again, within a short time, the population of flies increased, and they were again sprayed with Insecticide A. This sequence of events was repeated five times; then it became apparent that Insecticide A was becoming less and less effective in killing the flies.*

EXPLORE Solutions to the Problem. (15 mins.)

1. Have the student pairs discuss the problem and prepare several different explanations for the observations.

2. Have several pairs share their results with the class. Students might propose explanations similar to the following:

   • As Insecticide A aged, it becomes less effective.

   • Insecticide A is effective only under certain environmental conditions (e.g., a range of temperatures and humidity), and one or more of these environmental conditions changed.

   • Some flies survived and reproduced. (This item should not be elicited or developed at this point if suggested.)

3. Continue the inquiry with the story line. To the students: *The farmer related that the same insecticide solution had been made and used in all the sprayings. He thought the insecticide solution may have decomposed with age.*

4. Have the student pairs suggest a way to test this hypothesis. The students may propose several different investigations. One example would be to use sprays of different ages on the flies and determine which sprays were most effective. Students may suggest making a chemical analysis of new and old solutions to determine if changes had occurred.

5. To the students: *Researchers made a new batch of Insecticide A. They used it instead of the old batch on the fly population. The result was that only a few of the flies died. The same batch of the insecticide was then tried on a different fly population in a distant barn. In this case, the results were like the original spraying; most of the flies were killed. Here were two quite different results with a fresh batch of Insecticide A. Moreover, the weather conditions at the time of the effective spraying of the distant barn were the same as when the spray was used without success at the first barn.*

6. Have the student pairs analyze the observations and list the major components of the problem and subsequent explanations. Students should list what they know, what they propose as explanations, and what they could do to test their explanations. Students might list the following:

   • Something about the insecticide itself

   • The conditions under which the insecticide was used

   • The way in which the insecticide was used

   • Something about the organisms on which the insecticide was sprayed

7. To the students: *So far, your explanations have had to do with just a few of these components. Which ones?*

8. The explanations have concerned only "something about the insecticide" and "the condition under which the insecticide was used."

9. To the students: *The advantage of analyzing a problem as we have done is that we can identify the possibilities we have not considered. What possibilities in the list have we not considered in forming our hypotheses?* Direct students' attention to "something about the organisms on which the insecticide was sprayed."

EXPLAIN Natural Selection. (15 mins.)

1. At this point you should incorporate an initial explanation of natural selection. The explanation should be based on the prior discussion and analysis of the fly extermination problem.

2. Recall that natural selection occurs because individual organisms vary in their traits. Some of those variations can contribute to an individual's survival and reproductive advantage. In time, those individuals that survived and reproduced contribute more offspring to the population. This increase in the frequency of favorable traits results in increased adaptability to the environment. This is what happened as spraying continued on the population of flies in this invitation to inquiry.

3. To the students: *Let us examine the interactions between Insecticide A and the flies. Applying your knowledge of biology, what might have happened within the fly population that would account for the decreasing effectiveness of Insecticide A?*

4. Ask the students to remember that after the first spraying, most, but not all, of the flies did not survive. Ask them where the new population of flies came from, or who were the parents of the next generation of flies? Were the parents among the flies that did or did not survive the effects of Insecticide A? Remind the students that the barn was sprayed a second time. If there are differences in the population to Insecticide A susceptibility, which individuals would be more likely to survive this spraying? The students might thus be led to see that an explanation based on the idea of natural selection, in this case in an insecticide-dominated environment, might have resulted in the survival of only those individuals that were best adapted to live in that environment. The insecticide did not kill all of the flies. Those flies that survived passed on the capacity to survive subsequent sprayings. The frequency of the ability to survive insecticide increases in the population.

ELABORATE the Idea of Natural Selection. (15 mins. 11 secs. for DVD segment)

1. Show the *EVO* DVD Question 3, What is natural selection? This segment of *EVO* will expand students' understanding of natural selection.

2. After the clip answer any questions students might have. You can use the *Look-Again* markers to return to the following clips:
   - spider and eggs, more are born …
   - robins with babies

- acorns to oak trees
- what Darwin discovered, mimicry examples
- brown snails in yellow grass animation
- natural selection defined
- Profs. Grant and Darwin finches
- blue-footed boobies, sexual selection
- cock-of-the-rock
- Harris hawk supplanting

EVALUATE Students' Learning.

1. Immediately after class (or school), reflect on students' understanding of natural selection, their ability to develop explanations, and their skills at problem solving.
2. Complete the evaluation for the student pairs that worked together.
3. Use the following criteria for evaluating students' learning:

   Exceeds expectations: All outcomes are met, work goes beyond expectations and demonstrates exceptional understanding.

   Meets expectations: Work indicates that student understands major issues and concepts.

   Below expectations: Student work does not meet the criteria. It may be incomplete or may not address key issues or concepts of the learning outcomes.

Learning Outcomes	Evaluation
Develop knowledge about natural selection	
Develop ability to create explanations based on evidence	
Develop problem-solving skills	
Develop communication skills	
Overall evaluation for this lesson	

LESSON FOUR:
How Do Species Come About?

OVERVIEW

This lesson begins by showing the first section of Question 4 of the *EVO* DVD. Students are first engaged by the question, What is a species? The lesson continues with a discussion of the concept of species. Students are confronted with the problem of the origin of new species and provided with an explanation using the *EVO* DVD. In the final portion of the lesson, students use treelike diagrams to represent the emergence of new species (i.e., descent from common ancestors). **Note:** This lesson is best used as an introduction, a summary, or a review and should be complemented by further study of speciation. It only briefly presents ideas that will require more extensive study.

Time: One 50- to 55-minute class period   **DVD: 11 mins. 1 sec.**

LEARNING OUTCOMES

As a result of this lesson, students should be able to

- learn the concept of species,
- understand that species descend from common ancestors,
- understand the process of speciation, and
- develop adaptability and complex communication skills.

MATERIALS FOR THE LESSON

- Equipment to play the *EVO* DVD for the entire class and computer workstations for students to view the *EVO* DVD during group classwork. Arrange groups of four to five students each.

BACKGROUND FOR THE TEACHER

This lesson on speciation presents ideas central to understanding biological evolution. Portions of the following explanation have been adapted from *The Nature of Science and the Study of Biological Evolution* (BSCS 2005).

The lesson begins with the concept of species. Students have a general understanding of what species are. They know, for example, that an evergreen and an oak tree are two different types (i.e., species) of trees. However, scientists have found it difficult to form a single definition of species that can be used universally.

Darwin described a key characteristic of a species as members of the same species can interbreed in nature and produce fertile offspring. In the 20th century, biologist Ernst Mayr used interbreeding in a formal definition called the *biological species concept*. It defines a species as any population or group of populations whose members have the potential to interbreed with one another in nature to produce viable, fertile offspring. Organisms that cannot interbreed in nature to produce such offspring are said to be *reproductively isolated* and belong to different species.

This definition focuses on interbreeding rather than the extent to which organisms resemble one another. It is a useful concept, because some organisms look alike but never interbreed in nature. They, therefore, belong to different species. On the other hand, some organisms look different but interbreed, so they are members of the same species. Because all humans can mate and produce fertile offspring, all humans are members of the same species *Homo sapiens*.

Since members of different species do not interbreed, the genetic makeup (or gene pool) of each species remains separate from other species. In genetic terms, there is no gene flow, or exchange of genes, between species. Because species are reproductively isolated in this way, each evolves without genetic input from other species and is likely to acquire different variations across many generations.

You may want to point out to students the limitations of the species concept and have students think about how to solve problems associated with the limitations. For example, scientists cannot apply it to fossil organisms or those that reproduce asexually, such as bacteria. In these cases, scientists rely on appearance. There also are situations in which scientists do not know whether particular organisms can or would interbreed in nature because they live in different habitats. Lions, for example, live in open grasslands, and tigers live in forests. They live apart and do not interbreed in nature. However, lions and tigers will mate and produce fertile offspring when brought together in a zoo. Since lions and tigers have at least the potential to breed in the wild, the biological species concept has limitations in this case.

To solve such problems, scientists have proposed another definition of species: the *ecological species concept*. It defines a species in terms of its ecological niche (i.e., its habitat), the environmental resources it uses, and its behaviors. Under this definition, lions and tigers are considered separate species, because their habitats and behaviors, not reproductive incompatibility, keep them from interbreeding. The important point is that the basic evolutionary relatedness of all life and the constant slow change all populations undergo mean that every definition of species is bound to have limitations.

Another alternative to the species question is to consider species as lineages that exist through time. This is a perspective suggested by Kevin de Queiroz in the *EVO* DVD.

How do species come about? Scientists explain speciation as a result of temporary isolation of a population from its parental population. During isolation, the genetic makeup of the two populations changes independently of one another due to mutation, genetic drift, and natural selection. Eventually, the two populations become so different that they cannot successfully interbreed. Reproductive isolation develops when two or more populations of a single species become geographically separated from one another and begin to change. The term for this process is *geographic speciation*. The Galápagos mockingbird examined by Darwin and reviewed in Question 2, Who was Charles Darwin? is an example of geographic speciation (See the mockingbird speciation animation *Look-Again* marker in Question 2 of the *EVO* DVD). During geographic speciation, a population of one species becomes subdivided into two or more groups by some type of geographic barrier, such as a river, a canyon, or an expanse of ocean. The barrier prevents individuals in one group from interbreeding with those of the other. This isolation prevents the groups from mixing genes. Across time, each population might change in response to its environment differently from the other populations so that the populations become different.

For instance, natural selection might modify one group of organisms in one direction and another group in another direction. The groups may eventually come to differ so much physically or behaviorally that they do not interbreed even if the barrier is removed and the now-modified groups have contact with each other. If this happens, scientists call each group a different species.

Evolutionary divergence between populations goes through a series of stages with time. For example, if an ancestral population splits into two isolated local populations, these populations may diverge genetically, physically, and behaviorally across time and become two subspecies, then two species, then perhaps even two genera. Genetic data support this pattern of evolutionary change.

Speciation can also occur in other ways. It sometimes occurs when no geographic barriers separate populations, such as an ecological environment favoring individuals at the extreme ends of genetic variation. When this happens, natural selection can result in two distinct groups with different physical characteristics. Across time, these groups may become separate species.

The evolution of many diverse species from a common ancestor species is called *adaptive radiation*. Examples of adaptive radiation include the evolution of 13 species of finches on the Galápagos Islands and more than 500 species of drosophilid flies (relatives of the fruit fly *Drosophila*) on the mountainsides of Hawaii. In both cases, geographic separation (different species found on different islands) and ecological specialization (different species found at different elevations on the same mountainside) played a role in speciation. These examples show that speciation is a process of repeated branching and diversification of new species from ancestral species.

Note: This DVD segment takes a stance against human exceptionalism. This may be a source of discontent for some students. Students' concerns should not be suppressed or denigrated, rather make the point that science places value on all species.

PREPARING FOR THE LESSON

- The lesson involves extensive use of the *EVO* DVD, in particular, the use of the *Look-Again* markers to progress through the lesson. Due to this use of the *Look-Again* markers, a preliminary review of *EVO* and the coordination of phases of the lesson are strongly recommended. The *Look-Again* markers will direct viewers to the following clips:
    - species defined
    - phylogenetic trees
    - common ancestor defined
    - man and chimps
    - speciation
- Review *EVO* Question 4, How do species come about?

LEADING THE LESSON

ENGAGE the Students. (10 mins.)

1. Ask the students if they have heard the word *species*? Where did they hear the word? What does it mean?

2. Tell them that this lesson answers the question, How do species come about? But first students have to understand what a species is.

3. Go to the *EVO* DVD Main Menu and click on Extras. Then click on *Look-Again* and go to the marker species defined for Question 4, How do species come about?

EXPLORE the Origin of Species. (10 mins.)

1. Play the section on species defined. The section continues through a discussion of speciation by Niles Eldredge.

2. Ask the students if they have any questions and take some time to provide clarification of factors that contribute to how new species come about.

EXPLAIN the Tree of Life. (10 mins.)

1. Go to the *Look-Again* marker phylogenetic trees and play this segment of the *EVO* DVD.

2. Let the clip continue until the end.

3. Ask students for questions and provide an explanation of phylogenetic trees, common ancestors, scala natura, and a final summary of speciation.

4. Ask the students to explain the difference between a "ladder," or hierarchy form and a phylogenetic tree.

ELABORATE the Idea of Speciation. (20 mins.)

1. Give the students the following information. To the students: *On the Galápagos Islands, there are 13 different species of finches. Some have big beaks, others small beaks, and the different species eat different things.*

2. Have the students work in groups of four or five to explain how the different species may have evolved. Student groups may have to do some background research on the Galápagos Islands to identify environmental and resource characteristics that may have influenced evolution of new species. The explanation should include the general observations of geographic separation (different species on different islands) and ecological specialization (different species in different ecological niches).

EVALUATE Students' Learning.

1. Give student groups the following information:

   Darwin's Dreampond: Drama in Lake Victoria *by Tijs Goldschmidt (1996) tells the biological story of cichlid fishes in Lake Victoria in East Africa. They present an example of speciation and adaptive radiation (i.e., Diversification into different ecological niches by species derived from a common ancestor). Lake Victoria is a large lake that is less than a million years old. In the history of Lake Victoria, nearly 200 cichlid species have evolved in this body of water. Because all of these fishes are closely related genetically, it is likely that they all evolved from one or a few ancestor species that first entered the lake during formation. Provide a scientific explanation that may account for the evolution of new species in Lake Victoria.*

   Sample summary response: New species evolve from existing species through the process of speciation and adaptive radiation. Speciation often involves populations becoming geographically isolated from other populations and evolving independently of each other under the influence of natural selection and other evolutionary forces.

   In some cases, new species evolve through ecological isolation and the need to adapt in order to utilize the available resources for survival and reproduction. In the lake there are different environments and temperatures at different depths and different plants, animals, and other matter.

2. Remind students that this is a team effort. Students will hand in one summary statement.

3. Assessment of student learning should be judged on the basis of the summary statements.

4. Use the following criteria for evaluating students' learning:

   Exceeds expectations: All outcomes are met, work goes beyond expectations and demonstrates exceptional understanding.

   Meets expectations: Work indicates that student understands major issues and concepts.

   Below expectations: Student work does not meet the criteria. It may be incomplete or may not address key issues or concepts of the learning outcomes.

Learning Outcomes	Evaluation
Learn the concept of species	
Understand that species descend from common ancestors	
Understand the process of speciation	
Develop adaptability	
Develop complex communication skills	
Overall evaluation for this lesson	

LESSON FIVE:
Where Do Variations Come From?

OVERVIEW

This lesson begins with students observing genetic variations within a population they know well: their own populations. They confront the question, How can you explain the variations? The *EVO* DVD provides an overview of the origin of genetic variations. After further review and clarifying explanations, the idea of symbiogenesis is introduced. **Note:** This lesson is best used as an introduction, summary, or review and should be complemented by further study of genetics.

Time: One 50- to 55-minute class period   **DVD: 9 mins. 40 secs.**

LEARNING OUTCOMES

As a result of this lesson, students should be able to

- understand the origins of genetic variation,
- know and use gene flow, genetic drift, and genetic mutations as the variations that serve as the basis for natural selection,
- understand symbiogenesis,
- develop systems thinking, and
- develop communication skills.

MATERIALS FOR THE LESSON

- Equipment to play the *EVO* DVD for the entire class

BACKGROUND FOR THE TEACHER

This lesson on genetic variation presents ideas central to biological evolution. It is best used as a complement (i.e., an introduction, summary, or review) for more extensive study of genetics. Portions of the following explanation have been adapted from *The Nature of Science and the Study of Biological Evolution* (BSCS 2005).

Evolution can be defined as a change through geological time in the genetic composition of a population. The genetic makeups of populations change at different ages, at different times. The factors causing these changes include natural selection, gene flow, genetic drift, mutations, and symbiogenesis. These factors act continuously. The result of the continual actions of these factors means that every population is constantly evolving. The gene pool of a population is always changing. The result is a change in the population's observable (i.e., phenotypic) traits. While natural selection is certainly important, researchers have not resolved just how important it is in all circumstances. Other evolutionary factors, such as gene flow, genetic drift, and mutations, contribute to the evolution of populations.

Natural selection occurs because individuals vary in the degree to which they can survive and reproduce in a given environment. Individuals whose genetic traits make them better adapted to an environment tend to contribute more offspring to future generations than individuals who are less well adapted. Better-adapted individuals pass traits with a survival advantage to their offspring through inheritance. Both the distribution of traits and the gene frequencies in populations change, and traits with adaptive value and a survival advantage become more frequent.

All individuals of sexually reproducing species vary in their physical, physiological, biochemical, and behavioral traits. For example, plants vary in flower color, petal shape and size, leaf size, and many other physical traits. Much of the variation in populations is *continuous*. That is, there are small differences among individuals.

Genetic recombination during meiosis is an important source of variation in populations of sexually reproducing species. In genetic recombination, each individual's gametes (i.e., sex cells such as eggs or sperm) contain different combinations of their alleles (i.e., one of several forms of the same gene, some differing due to a mutation of the DNA sequence). When the gametes from different individuals combine, the offspring receives half of its alleles from each parent. As a result, the offspring of two parents will be different from each other genetically (unless they are identical twins).

Gene flow also influences the distribution of variation in a population. The genetic makeup of a population changes if individuals move into or out of the population. Individuals moving into a population bring their alleles into the population's gene pool; individuals moving out remove their alleles from the gene pool. Gene flow also takes place when pollen grains, spores, or gametes move between populations. The flow of alleles into or out of a population changes the population's gene pool. Mathematics can be used to measure the effect of gene flow. Scientists have discovered that two factors determine how much the gene pool of a population will change because of gene flow: (1) the number of individuals moving into and out of a population across time and (2) allele frequency differences between the migrants and the population.

Genetic drift refers to random changes in allele frequencies due to chance. These random changes sometimes result in alleles disappearing from a population. The smaller the population, the more likely the gene pool will change significantly due to genetic drift. Rarely does genetic drift result in phenotypic change.

Genetic mutations also contribute to variation. A mutation is a spontaneous change in an organism's DNA. Mutations are caused by DNA copying errors and by environmental factors, such as radiation. Point mutations are molecular changes in the nucleotide base sequence of a gene. Such mutations can alter an organism's observable traits (i.e., its phenotype).

Mutations can occur in the DNA of any cell of your body. Most cells do not pass on genetic information to the next generation. However, if a mutation occurs in a cell that produces eggs or sperm, then you can pass this mutant allele to your offspring.

Mutations are not usually a major evolutionary force for changing the genetic makeup of a population from one generation to the next. Mutations alone do not bring about big changes in a population's gene pool. Furthermore, most mutations have negative effects, so they are likely to be eliminated from the population. However, mutation is important as a source of new variation in a population. Mutations can produce new alleles that recombine with other alleles during meiosis, which are then acted upon by natural selection.

Another source of variation is *symbiogenesis*, which accounts for significant changes. For example, mitochondria and chloroplasts that originated as bacteria lived as internal symbionts of early eukaryotic cells. The modern explanation—symbiogenesis—is attributed to Lynn Margulis.

PREPARING FOR THE LESSON

- The lesson involves fundamental genetics (e.g., gene flow), mutations that account for genetic variation in a population. It is recommended that in addition to teacher background, you review genetic fundamentals prior to the lesson.

- Review *EVO* Question 5, Where do variations come from?

LEADING THE LESSON

ENGAGE Students in the Idea of Genetic Variation. (10 mins.)

1. Begin the lesson by introducing a simple characteristic such as height, natural hair color, or eye color. Be sure to use an inherited characteristic. Have students observe the variations within the characteristic in the classroom.

2. Once they have observed the variations within the class, ask them how the variations came about. They will likely answer that they inherited the characteristic from their parents.

EXPLORE the Role of Variations in a Population. (10 mins.)

1. Ask the students how variations come about in populations. The point is to have students begin thinking about population genetics, changes in traits, and the distribution of a trait, such as height, in a population.

EXPLAIN Genetic Variation in a Population. (9 mins. 40 secs. for the DVD segment)

1. Show the *EVO* DVD Question 5, Where do variations come from?

2. After showing the DVD take time to answer students' questions. You may find it necessary to use the *Look-Again* markers. The *Look-Again* markers will direct viewers to the following clips:
   - intro to genetics, four gen. photo
   - DNA, chromosomes, mitosis video
   - mutations defined
   - sexual recombination
   - wood frogs, eggs, tadpoles
   - symbiogenesis, Lynn Margulis
   - endosymbiosis explained

ELABORATE Students' Understanding of Genetic Variations and Evolution. (25 mins.)

1. Present students with the following situation. To the students: *Cypress trees in southern California grow in small, isolated groves. The trees in each grove look similar to one another but quite different from trees in other groves. How would you explain these differences?*

Sample response: Scientists have discovered that these isolated populations of cypress trees in California are different due to genetic drift. Genetic drift refers to random changes in allele frequencies due to chance. These random changes sometimes result in alleles disappearing from a population.

EVALUATE Students' Learning.

1. Provide students with a new problem, one that requires an understanding of genetic variation and natural selection. The following situation may be used: Scientists have observed that nearly all water snakes (*Nerodia sipedon*) living along the shoreline of western Lake Erie have a dark-banded color pattern. Most of them living on islands within the lake are lighter colored and have reduced or no banding.

2. Ask the students to provide a scientific explanation that includes genetic variation as applied to evolution.

   Sample summary response: Evolution is a change in the genetic makeup of a population. Factors influencing changes in the genetic makeup of a population include natural selection, gene flow, genetic drift, and mutations. Although all four factors may be acting on the variation of traits in a population, a proposed explanation centers on natural selection and states that unbanded snakes living on the light-colored rocks of island shorelines are better camouflaged against predators than banded snakes. So, unbanded snakes in this environment have a higher survival rate. Banded snakes, on the other hand, are better camouflaged in the dense marshland vegetation that grows along the shoreline of the lake.

3. Assign this evaluation as homework, and have students work individually.

4. Assessment of student learning should be judged on the basis of the summary.

5. Use the following criteria for evaluating students' learning:

   Exceeds expectations: All outcomes are met, work goes beyond expectations and demonstrates exceptional understanding.

   Meets expectations: Work indicates that student understands major issues and concepts.

   Below expectations: Student work does not meet the criteria. It may be incomplete or may not address key issues or concepts of the learning outcomes.

Learning Outcomes	Evaluation
Understand the origins of genetic variation	
Know and use gene flow, genetic drift, and genetic mutations that serve as the basis for the process of natural selection	
Understand symbiogenesis	
Develop systems thinking	
Develop communication skills	
Overall evaluation for this lesson	

LESSON SIX:
What Role Does Cooperation Play?

OVERVIEW

This lesson begins by asking students to think of a situation where they <u>both</u> compete and cooperate. They then identify a team sport such as basketball or volleyball to explore the theme of cooperation. The *EVO* DVD explains that cooperation has an important role in evolution. The DVD also provides numerous examples of cooperation and contrasts it with competition. Student groups use cooperation to analyze an example of cooperation in nature.

Time: One 50- to 55-minute class period   **DVD: 9 mins. 43 secs.**

LEARNING OUTCOMES

As a result of this lesson, students should be able to
- identify cooperation as a major factor of evolution,
- recognize the role of cooperation in animal and human societies,
- demonstrate cooperation in a group of peers, and
- analyze the adaptability achieved via cooperation.

MATERIALS FOR THE LESSON

- Equipment to play the *EVO* DVD for the entire class

BACKGROUND FOR THE TEACHER

In very simple terms, one can use *roles* and *goals* as a way of thinking about and analyzing cooperative interactions. Think of a team sport such as basketball. Each player has a role and the team cooperates to achieve a goal: to win the game.

In discussions of evolutionary biology, competition among species has historically been emphasized more than cooperation. Competition, for example, may take the form of one population consuming a resource that is then not available to another population. The second type of competition is when two individuals directly interact and one loses the encounter. The latter is by far the most common example of competition in discussions of biological evolution.

In contrast to competition, there are many beneficial interactions among species. One such interaction is *commensalism*. Commensalism is a relationship in which one species benefits from another species, while the latter species is not affected by the interaction. Dispersal of seeds on the fur of animals is an example of commensalism. *Mutualism* is another beneficial interaction among species. In this case, each species exploits each other as resources so both benefit from the interaction. A third type of relationship is *parasitism*. Here one species benefits while harming the other species. Ticks, for example, are parasitic organisms.

In terms of biological or evolutionary fitness, interactions among individuals may have four possible results. One result is *mutualism* (or cooperation), the outcome of which is increased fitness for both species. *Altruism* represents a second possible interaction. Here, the individual initiating the action has a fitness cost, and the organism receiving the action benefits. A third result is *selfishness*, which is the opposite of altruism. The individual initiating the action benefits, and the organism receiving the action loses. Finally, there may be behaviors that result in fitness losses for both organisms. This interaction is referred to as *spite*. Scientists cannot find evidence for this interaction in nature. The reason that spite cannot be found in nature seems clear: An allele that does not contribute to fitness by either the actor or recipient would be eliminated due to natural selection. It has no survival value for either organism.

The *EVO* DVD presents cooperation in the following way. Cooperation occurs between

- levels of organization within the same individual
  - cells, organs, systems
- individuals within a species
  - ants and other social insects
  - humans
  - culture
- individuals from different species

Cooperation between species also is regarded as one type of symbiotic relationship.

When investigating the role of cooperation in evolution, Michod and other biologists would ask several questions: What is an individual? On what unit does natural selection act? A cell? An organ? A system? A reproductive couple? A family? This leads to the idea of group selection.

All direct interactions between the life cycles of individuals within different species are called symbiotic relationships. So, for example, some interactions are mutually beneficial (i.e., mutualism and cooperation), some interactions benefit one and do not have much effect on the other (i.e., commensalism), and some interactions benefit one and harm the other (i.e., parasitism).

Two phases of this lesson—elaborate and evaluate—require students to work in cooperative groups. For the group work, the goal should be clearly stated. The team members should each be accountable for the results. For example, all members should be able to explain how the team solved the problem and what the proposed solution is. The various roles for the team should be clear. Leadership for the group is shared as each member has a task. Your role as the teacher is to serve as a consultant, providing directions, suggestions, feedback, and advice.

PREPARING FOR THE LESSON

- Review *EVO* Question 6, What role does cooperation play?

LEADING THE LESSON

ENGAGE Students in the Ideas of Cooperation and Competition. (5 mins.)

1. To the students: *Can you think of a situation where you <u>both</u> compete and cooperate?* The likely responses will be team sports such as basketball, volleyball, softball, or baseball.

EXPLORE the Idea of Cooperation. (10 mins.)

1. Have the students identify a team sport (be sure the sport is played by both girls and boys) such as basketball.

2. Ask the students to identify elements of cooperation and competition in the sport. They might identify winning against another team (competition), having a role as an individual player, and running team plays (cooperation).

3. Tell the students that cooperation plays a role in biological evolution and they are going to study cooperation among organisms.

4. Ask the students to try and identify an example of cooperation between organisms of the same species and between different species.

EXPLAIN Cooperation as a Force in Biological Evolution. (9 mins. 43 secs. for DVD segment)

1. Show the *EVO* DVD Question 6, What role does cooperation play?

2. Answer students' questions about the DVD segment. You may use the *Look-Again* markers to clarify the discussion. *Look-Again* markers will direct viewers to the following clips:

   - ants saving eggs
   - cooperation and conflict
   - cooperation defined, group selection
   - ants and aphids, symbiosis defined
   - human zoo
   - system of Earth, Gaia

ELABORATE the Role of Cooperation in Biological Evolution. (25 mins.)

1. Have the students work in groups of four. They are to review the DVD Question 6 and appropriately apply the concepts of mutualism, altruism, commensalism, selfish behaviors, and symbiosis.

2. Emphasize that students will cooperate in working on this task. Help them begin by indicating they should establish a goal for the group work (i.e., analyze clips of the DVD and prepare a report) and roles for team members (i.e., summarize observations, contribute to analysis, prepare report, obtain resources, and conduct research).

3. Specifically, the teams should use an evolutionary perspective to provide an explanation for the following:

   - ants and aphids
   - ants saving eggs
   - clown fish and anemone

EVALUATE Students' Learning.

1. It would be possible to use the group work on the prior task as the content for this evaluation.

2. If you wish to provide a separate evaluation, you should structure the same groups as formed for the elaborate stage of the lesson, because they will have had time to establish norms of behavior and working relationships. Students could analyze the following situation for the evaluation:

   Two scientists observe two American crows (*Corvus brachyrhynchos*) patrolling the parameters of their respective nesting territories. One crow crosses the boundary, and the other initiates aggressive calls and gives chase, and the two crows engage in a fight. Later, a hawk flies over their territory, and both crows chase the hawk away. As the scientists continue observing the crows, they note that these two crows spend time getting food and feeding all the young birds in their nests, not only their own offspring.

3. Based on their understanding of cooperation, fitness, and evolution in general, the student groups should propose an explanation for the crows' behavior.

4. Assessment of students' learning should be based on (1) their reports and (2) your observations of their adaptability (cooperation) as a group. You should provide one assessment for each group.

5. Use the following criteria for evaluating students' learning:

   Exceeds expectations: All outcomes are met, work goes beyond expectations and demonstrates exceptional understanding.

   Meets expectations: Work indicates that student understands major issues and concepts.

   Below expectations: Student work does not meet the criteria. It may be incomplete or may not address key issues or concepts of the learning outcomes.

Learning Outcomes	Evaluation
Identify cooperation as a major factor of evolution	
Recognize the role of cooperation in animal and human societies	
Demonstrate cooperation in a group of peers	
Analyze the adaptability achieved via cooperation	
Overall evaluation for this lesson	

LESSON SEVEN:
What Is a Brief History of Life?

OVERVIEW

This lesson helps students understand the scale of geologic time and the evolution of life. Students try to figure out something within their understanding that equals one million. They are then asked to complete a scale model of geologic time. After the *EVO* DVD outlines the evolution of life, students return to the model of geologic time and insert events in the evolution of life.

Time: Two 50- to 55-minute class periods   **DVD: 8 mins. 45 secs.**

LEARNING OUTCOMES

As a result of this lesson, students should be able to
- understand the scale of geologic time,
- know the evolution of life in the scale of geologic time,
- communicate the scale of geologic time and evolution of life,
- develop problem-solving skills, and
- apply systems thinking.

MATERIALS FOR THE LESSON

- 5 m long strips of paper, one per pair of students (Shelf paper may be cut into strips for the student pairs.)
- One meterstick per pair
- Copies of Figure 2, A Geologic Time Scale (p. 42), and Figure 3, A Time Scale for the Evolution of Life (p. 43), for each student pair
- Equipment to play the *EVO* DVD for the entire class and computer workstations for students to view the *EVO* DVD during group classwork. Arrange groups of four to five students each.

BACKGROUND FOR THE TEACHER

Most students have problems with the scale of geologic and evolutionary time. In this lesson, students use the following scale:

1 millimeter = 1 million years
1 meter = 1 billion years

Students will struggle with placing events within 1 million years, making it difficult for them to understand the scale of time involved in explaining the evolution of life. You should let them struggle with the problem but not to the point of rejecting the task.

Geologic time is subdivided on the basis of the amount and type of activity within Earth systems and the evolution of life. Geologic time is ordered both relatively and absolutely. For relative dating, the sequence in which rock strata formed is used. Fossils also provide guides to Earth's history as scientists assigned relative dates to the world's rocks according to fossil evidence.

Radiometric dating provides absolute ages for events in Earth's history. Radiometric dating techniques apply the decay rates of selected naturally radioactive isotopes to stable daughter isotopes to determine how long the unstable parent isotopes have been decaying. Fairly accurate dates have been determined for the events beginning in the Cambrian era; this comprises about 12% of the Earth's history.

PREPARING FOR THE LESSON

- Prepare the strips of paper (approximately 3 cm wide and 5 m long). Distribute one strip of paper for each pair of students.

- Distribute one meterstick for each pair of students.

- Make copies of Figure 2, A Geologic Time Scale ,and Figure 3, A Time Scale for the Evolution of Life.

- Review the *EVO* DVD Question 7, What is a brief history of life?

LEADING THE LESSON

ENGAGE the Students in the Scale of Geologic Time. (10 mins.)

1. Ask the students if they can identify something that is one million. For example, ask students the following questions. To the students: *Does the school or park lawn have a million blades of grass? How could you find out? When was a million seconds ago? A billion seconds ago?*

2. Then ask the students about the age of the Earth. To the students: *What's the most widely accepted current estimate of the age of the Earth?*

3. It is OK if students use handheld calculators or look for the answers using their computers.

EXPLORE the Scale of Geologic Time. (30 mins.)

1. Place students in pairs and tell them they are going to set up a scale for geologic time. Give them copies of Figure 2, A Geologic Time Scale, the 5m strips of paper, and the metersticks.

2. Students likely will need help in setting up the scale. Give them the suggestions that a reasonable scale is 1 mm to 1 million years, 1 cm to 10 million years, and 1 m to 1 billion years. (**Note**: Regardless of the scale students select, the most recent time—about one million years—will present a problem. An important learning outcome about the scale of time will develop when students work out and have to solve this problem.)

3. Allow students to work out the scale with your review, questions, and suggestions.

4. Students are to record the information from Figure 2, A Geologic Time Scale on their paper strips.

EXPLAIN Geologic Time and the Evolution of Life. (8 mins. 45 secs. for DVD segment)

1. Show the *EVO* DVD Question 7, What is a brief history of life?

2. Answer any questions students have about the clip. You may use *Look-Again* markers to help answer student questions. *Look-Again* markers will direct viewers to the following:
   - origin of life, bacteria fossils
   - nucleus, mitochondria, chloroplasts
   - Cambrian explosion
   - age of human species
   - punctuated equilibria, Niles Eldredge
3. Explain the need to grasp the scale of geologic time to understand the process of biological evolution.

ELABORATE Students' Understanding of Geologic Time and the Evolution of Life. (50 mins.)

1. Present the students with Figure 3, A Time Scale for the Evolution of Life.
2. Tell the students to incorporate the information on A Time Scale for the Evolution of Life into their scale of geologic time. (**Note:** The students may have trouble with the scale of most recent time, approximately a million years.)
3. Provide guidance and clarifying questions to help students complete the scale.
4. Hold a concluding discussion on the history of life. Use the *EVO* arm graphic to give students a personal perspective. You may need to return to the *EVO* DVD.

EVALUATE Students' Learning.

1. Ask each student pair to write a brief summary explaining the time scales and evolutionary changes. They should use evidence from fossils and radiometric dating to support their summaries.
2. Assessment of students' learning should be judged on the basis of your observation of the classwork and final summary.
3. Use the following criteria for evaluating students' learning:

   Exceeds expectations: All outcomes are met, work goes beyond expectations and demonstrates exceptional understanding.

   Meets expectations: Work indicates that student understands major issues and concepts.

   Below expectations: Student work does not meet the criteria. It may be incomplete or may not address key issues or concepts of the learning outcomes.

Learning Outcomes	Evaluation
Understand the scale of geologic time	
Know the evolution of life within the scale of geologic time	
Communicate the time scale for biological evolution	
Develop problem-solving skills	
Apply systems thinking	
Overall evaluation for this lesson	

Figure 2. A Geologic Time Scale

Formation of the Earth—4.6 billion years ago (4,600 million)

Pre-Cambrian, oldest rocks—4.0 billion years ago (4,000 million)

Pre-Cambrian, formation of super continent—1.5 to 1.0 billion years ago (1,000 million)

Beginning of the Cambrian—570 million years ago

Beginning of the Ordovician—520 million years ago

Beginning of the Silurian—450 million years ago

Beginning of the Devonian—420 million years ago

Beginning of the Mississippian—375 million years ago

Beginning of the Pennsylvanian—305 million years ago

Beginning of the Permian—285 million years ago

Beginning of the Triassic—240 million years ago

Beginning of the Jurassic—195 million years ago

Beginning of the Cretaceous—135 million years ago

Beginning of the Paleocene—65 million years ago

Beginning of the Eocene—60 million years ago

Beginning of the Oligocene—35 million years ago

Beginning of the Miocene—25 million years ago

Beginning of Pliocene—5 million years ago

Beginning of the Pleistocene and ice ages—2.5 million years ago

Last ice age—10,000 years ago

Note: Based on the date of the activity, change the following dates to match the format of the events above (i.e., number of years ago, number of months ago, number of days ago, etc.)

Golden age in Greece—200 BC

Industrial Revolution in England—Mid 18th Century

Man lands on the moon—1969

Last New Year's Day

Today

Figure 3. A Time Scale for the Evolution of Life

Proposed origin of life—3.6 billion years ago (3,600 million)

Early algae and bacteria—3.5 billion years ago (3,500 million)

Photosynthetic bacteria—3.0 billion years ago (3,000 million)

First eukaryotes (cells with nucleus)—1.6 billion years ago (1,600 million)

First multicellular organisms—1.0 billion years ago (1,000 million)

Significant evolution of marine life, algae dominant—570 million years ago

First land plants, first land invertebrates—475 million years ago

First seed plants, first land vertebrates (amphibians)—400 million years ago

Widespread forests, early reptiles—350 million years ago

Gymnosperms dominant, first dinosaurs —230 million years ago

Age of dinosaurs, first birds—195 million years ago

First mammals—200 million years ago

First modern birds—150 million years ago

First flowering plants—130 million years ago

Spread of flowering plants, first elephants—35 million years ago

Appearance of hominids, appearance of first apes—3.5 million years ago

Appearance of *Homo erectus*—1.8 million years ago

Appearance of *Homo sapiens*—1.0 million years ago

LESSON EIGHT:
What Is the Controversy?

OVERVIEW

This lesson begins with students sharing with one another their views about evolution. The *EVO* DVD presents the controversy. Students then write one paragraph titled "Biological Evolution: What Is the Controversy?" The major aims of the lesson center on students developing the ability to express their understanding and beliefs about evolution and listen to other students' views in a clear and civil manner.

Time: One 50- to 55-minute class period   **DVD: 9 mins. 59 secs.**

LEARNING OUTCOMES

As a result of this lesson, students should be able to

- know there are scientific and nonscientific views on biological evolution,
- understand the differences between scientific explanations and personal opinions and beliefs,
- recognize and accept the views of other students, and
- develop the ability to listen and respond to other students in a civil manner.

MATERIALS FOR THE LESSON

- Equipment to play the *EVO* DVD for the entire class and computer workstations for students to view the *EVO* DVD during group classwork. Arrange groups of four to five students each.

BACKGROUND FOR THE TEACHER

The title of this *EVO* question signals a controversy. In this case, the controversy centers on some fundamentalist groups that have tried to eliminate biological evolution from science classrooms, while others have proposed introducing ideas such as "intelligent design" as an alternative explanation.

Rather than having students engage in debate and try to decide which is right—science or religion—this activity uses the scientists' explanations as the basis for civil discussion. As a science teacher, your role in this lesson is to emphasize the learning outcomes of recognizing scientific versus nonscientific perspectives, and having students use active listening to learn about other students' views.

Science endeavors to use evidence to provide explanations and predictions about the natural world, such as the movement of planets, daily weather changes in climate, and the evolution of life on Earth. Religion provides many individuals with meaning and purpose of human existence, proper relationships with others, and morals by which they should live. Other discussions of science and religion are in the references at the end of this section.

When asked about a definition of science, most science teachers express the complementary ideas that science is a body of knowledge and a process. Few would disagree with the late Carl Sagan when he stated in *Broca's Brain: Reflections on the Romance of Science* (1993), "Science is a way of thinking much more than it is a body of knowledge." The emergence of modern science in the late 16th and early 17th centuries was primarily due to the acceptance of new ways of thinking and explaining the natural world.

What are the basic elements of a scientific way of thinking? Briefly, a scientific explanation of nature must be based on empirical evidence from observations and experiments. Proposed explanations about how the world works must be tested against empirical evidence from nature. The scientific way of thinking stands in contrast to other ways of explaining nature (e.g., the acceptance of statements by authority, personal revelation, or religious doctrine). After the scientific revolution such approaches to explaining the natural world were no longer considered "scientific." They were not acceptable to the scientific community. Explanations had to be subject to confirmation by empirical evidence. For example, Galileo's observations of heavenly bodies confirmed Copernicus' heliocentric explanations of planetary motion. Since the emergence of modern science, our understanding of the natural world has progressed through consistent testing and the increasing appeal to empirical explanations. One could reasonably argue that the scientific way of knowing is among the great achievements of humankind.

Science teachers should review the following brief summaries about the nature of science and the ways of knowing prior to this lesson:

Science distinguishes itself from other ways of knowing and from other bodies of knowledge by using empirical standards, logical arguments, and skepticism as the criteria for possible explanations about the natural world.

Scientific explanations must meet certain criteria. First, they must be consistent with experimental and observational evidence about nature and must make accurate predictions, when appropriate, about systems being studied. They also should be logical, respect the rules of evidence, be open to criticism, report methods and procedures, and make knowledge public. Explanations of how the natural world changes based on other ways of knowing, beliefs, or explanations may be personally useful and socially relevant, but they are not scientific.

Because all scientific ideas depend on experimental and observational confirmation, all scientific knowledge is, in principle, subject to change as new evidence becomes available. The core ideas of science, such as the conservation of energy or the laws of motion, have been subjected to a wide variety of confirmations and are therefore unlikely to change in the areas in which they have been tested. (adapted from *The National Science Education Standards*, NRC 1996)

The National Science Education Standards and most state standards support the teaching of biological evolution.

Important professional organizations for science teachers, specifically the National Science Teachers Association (NSTA), the National Association of Biology Teachers (NABT), and Earth science teachers' associations have position statements that support the teaching of evolution. These position statements are complemented by those of scientific groups such as the National Academies, American Institute of Biological Sciences, American Geologic Institute, American Association for the Advancement of Science, and other discipline-specific groups.

PREPARING FOR THE LESSON

- The different perspectives students have may be subtle. Careful review of the *EVO* DVD prior to the lesson is essential.

LEADING THE LESSON

ENGAGE Students. (5 mins.)

1. Begin by asking students if they have heard any controversies about teaching biological evolution. If a student has, ask him or her to briefly explain. Accept the student's comments.
2. Tell the students that the lesson will help them understand more about evolution and the nature of science. In addition, they should learn other students' views.

EXPLORE Different Views on Evolution. (10 mins.)

1. Place students in pairs.
2. Tell the students they are to take five minutes each and explain to each other what they know about biological evolution.
3. During this conversation, they are to practice active listening and only ask clarifying questions.

EXPLAIN the Controversy About Evolution. (9 mins. 59 secs. for DVD segment)

1. Show the *EVO* DVD Question 8, What is the controversy?
2. After the *EVO* DVD, ask students if they have any questions or comments. You should clarify any questions about science and accept religious comments and views.
3. Help students understand and answer the lesson's question, What is the controversy? If necessary, use the *Look-Again* markers to return to sections of the DVD segment. *Look-Again* markers will direct viewers to the following clips:
   - evolution summary
   - evolution, a scary idea
   - science and religion
   - creationism, evolution in schools
   - Susan Epperson story
   - intelligent design

ELABORATE Students' Understanding of Evolution. (30 mins.)

1. Assign the pairs the task of preparing a joint statement titled "Biological Evolution: What Is the Controversy?"

2. The statement should take the form of a press release.

EVALUATE Students' Adaptability.

1. Due to the nature of this lesson, it may be best not to evaluate students.

2. If you do decide to evaluate students, the assessment should explicitly center on skills such as recognizing the scientific versus nonscientific perspectives of other students.

3. Student performance should be judged on the basis of their abilities.

4. Use the following criteria for evaluating students' adaptability:

   Exceeds expectations: All outcomes are met, work goes beyond expectations and demonstrates exceptional understanding.

   Meets expectations: Work indicates that student understands major issues and concepts.

   Below expectations: Student work does not meet the criteria. It may be incomplete or may not address key issues or concepts of the learning outcomes.

Learning Outcomes	Evaluation
Know there are different views of biological evolution	
Understand the differences between scientific explanations and personal opinions and beliefs	
Recognize and accept the views of other students	
Demonstrate the skills of adaptability: listening and responding to other students in a civil manner	
Overall evaluation for this lesson	

LESSON NINE:
Is Evolution Random?

OVERVIEW

This lesson begins with students observing and explaining patterns in nature. They confront the issue of explaining patterns in nature using natural processes or purpose. They conduct a simple activity of sorting gravel, sand, and ash and observe the patterns and physical process. Then, the students are asked the question, Is evolution random? They view the *EVO* DVD and learn about randomness, patterns, and purposeful explanations. Finally, they apply their understanding to a new problem presented in the final segments of the *EVO* DVD.

Time: One 50- to 55-minute class period   DVD: 8 mins. 4 secs.

LEARNING OUTCOMES

As a result of this lesson students should be able to

- recognize patterns in nature as the result of physical/natural processes,
- describe randomness as applied to genetic variations,
- explain selection as a natural process, and
- combine the randomness of genetic variation and the algorithms of natural selection as an answer to the question, Is evolution random?

MATERIALS FOR THE LESSON

- Equipment to play the *EVO* DVD for the entire class and computer workstations for students to view the *EVO* DVD during group classwork. Arrange groups of four to five students each.
- Several (three to four) pictures of patterns in nature (e.g., cross-sections of a nautilus, sunflower, leaves, frost crystals, honeybee combs, pebbles sorted on a beach, similar species of animals and plants)
- Each group of students will need the following:
    - One 25 ml graduated cylinder
    - A mixture of gravel, pebbles, sand, and ash
    - A source of water

BACKGROUND FOR THE TEACHER

This lesson centers on three ideas—randomness, patterns, and purpose—as they relate to natural phenomenon in general and biological evolution in particular. The *EVO* DVD presents explanations for genetic mutations; they are random. The DVD presents scientists explaining natural selection: It is not random, but natural selection lacks purpose. As these are relatively complex ideas, the lesson uses observable patterns in nature as a way to help students understand the process and answer the question, Is evolution random?

Simple patterns in nature can be explained in terms of natural processes. For example, as rivers reach lakes, they deposit their "load" (i.e., rocks and other material) based on mass, density, etc. Larger rocks are deposited before smaller pebbles, and there is a natural gradation, or sorting.

Complex patterns in nature may be more difficult to explain. The processes of biological evolution demand explanations that are more difficult to comprehend. Due to these processes' complexity, some people suggest a designer with a purpose. Purposefulness is a possible nonscientific explanation, but biological evolution can be explained without evoking purpose. Ever since Darwin, scientists have developed increasingly detailed scenarios for the development of all sorts of eyes based on comparing anatomy, paleontology, and comparative genomics.

Darwin's theory of evolution proposes two steps that result in the complexity of living organisms. First, there are random changes in the genetic makeup of individual organisms. Second, there is exposure to selective pressures or forces and only some individuals with some variations survive and reproduce. In time, the genetic variation in the population changes toward those characteristics with greater survival and reproductive advantages. These processes, in extended periods of time, produce unity and diversity in the living world. The cumulative process of selection is not purposeful. Yet to answer the lesson's question, evolution is not random. The lesson should help students understand that scientists can explain patterns such as evolution without appealing to a designer or ultimate purpose.

PREPARING FOR THE LESSON

- Review the *EVO* DVD. Go to the menu's Select A Question, and review Question 9, Is evolution random?
- Set up equipment (e.g., a computer or DVD player) for class presentation.
- Set up materials for the demonstration.

LEADING THE LESSON

ENGAGE the Students. (5 mins.)

1. Begin by showing the students actual examples of natural patterns. Alternatively, pictures of patterns in nature are appropriate. The examples might include hexagon-shaped cells of a honeybee hive and seashells.
2. Ask the students to identify other examples of patterns in nature.
3. Ask them how they would explain the formation of patterns in nature. Are the patterns random? The students are likely to provide a variety of definitions and explanations, some in nature and others not natural. As this is the Engage stage, accept all the students' suggestions.

EXPLORE Patterns, Purpose, and Randomness in Nature. (20 mins.)

1. Assign students to groups of four to five and distribute graduated cylinders (25 ml), gravel, sand, and ash to each group. Provide a source of water.
2. Have the students observe and describe the gravel, sand, and ash.
3. Ask them to predict what will happen if the mixture of gravel, sand, and ash is added to the cylinder filled with water. Students will likely say that the gravel will go to the bottom with sand and ash sorting out in that order.
4. Have them place the mixture of gravel, sand, and ash in the cylinder.
5. Have them explain the process of sorting that they observed. Students will explain the process using terms such as *density*, *gravity*, *friction*, or *buoyancy* as the reason the particles formed layers.

6. Conclude this phase of the lesson by asking if the pattern they observed was random? Was there a purpose? Help them with a clarification of these terms using the activity as the context for the exploration.

EXPLAIN Patterns, Purpose, and Randomness in Evolution. (20 mins.)

1. Ask the students Question 9 from the *EVO* DVD. To the students: *Is evolution random?* Accept their answers. Listen for their use of concepts and processes from prior lessons and activities.

2. Show the *EVO* DVD.

3. Ask the students to apply what they know and heard on the DVD and answer the question, Is evolution random?

4. Use the *Look-Again* markers to return to sections to help the students understand the relationship between the randomness of genetic mutations and the processes of natural selection. *Look-Again* will direct viewers to the following:
   • random defined
   • sorting algorithms
   • nat. selection not random, swan sailing
   • genetic drift, snails and cows animation

ELABORATE the Students' Explanations.

1. Use the *Look-Again* markers to show the section on genetic drift, snails and cows animation.

2. Ask the students to use the terms *random*, *natural selection*, and *patterns* to explain what the scientist Futuyma explained.

EVALUATE Students' Learning.

Use the following criteria for evaluating students' learning:

Exceeds expectations: All outcomes are met, work goes beyond expectations and demonstrates exceptional understanding.

Meets expectations: Work indicates that student understands major issues and concepts.

Below expectations: Student work does not meet the criteria. It may be incomplete or may not address key issues or concepts of the learning outcomes.

Learning Outcomes	Evaluation
Recognize patterns in nature as the result of physical/natural processes	
Describe randomness as applied to genetic variations	
Explain selection as a natural process	
Combine the randomness of genetic variation and the algorithms of natural selection as an answer to the question, Is evolution random?	
Overall evaluation for this lesson	

LESSON TEN:
Why Should Anyone Care About Evolution?

OVERVIEW

This lesson serves as a culmination for *EVO*. The lesson begins with the question, Why should anyone care about evolution? Students present their ideas. The lesson continues to the *EVO* DVD where scientists briefly summarize their answers to the question. Students then work individually on short, two-page statements that represents their answers to the question.

Time: One 50- to 55-minute class period   **DVD: 11 mins. 36 secs.**

LEARNING OUTCOMES

As a result of this lesson, students should be able to

- synthesize knowledge about evolution,
- construct an argument using evidence,
- demonstrate communication and self-management skills, and
- demonstrate systems thinking.

MATERIALS FOR THE LESSON

- Equipment to play the *EVO* DVD for the entire class

BACKGROUND FOR THE TEACHER

This lesson is designed as a final evaluation. Students are asked to develop an argument for evolution. Using this approach makes connections to the language arts, in particular the Common Core Standards for Language Arts.

The approach used here also aligns with one aspect of scientific practice and inquiry—developing an argument using evidence. The practice of science requires reasoning, argument, and supporting one's position with evidence. The continuing challenge in science is to develop the best explanation for a natural phenomenon. The practice of science requires individuals to defend their explanations, bring evidence to support their positions, and to consider other evidence and explanations. This activity has parallels to these scientific practices.

In this activity, students will be required to (1) state a brief summary of evolution, (2) identify one example of a reason to understand and care about evolution, (3) include a quote from a scientist in the *EVO* DVD, and (4) cite two sources supporting the position taken in the short essay.

PREPARING FOR THE LESSON

- Review the *EVO* DVD. Specifically, review Question 10, Why should anyone care about evolution?
- Set up equipment (e.g., a computer or DVD player) for class presentation.
- Arrange for students to access computers as they work individually on their essays.

LEADING THE LESSON

ENGAGE the Students. (5 mins.)

1. Begin by asking the question, Why should anyone care about evolution?
2. Accept all answers as they set the stage for students' essays. If students need prompting, ask them to think back over the *EVO* DVD segments and other

lessons on evolution. What are the applications of evolution? Can they think of life situations in which understanding evolution might be helpful?

EXPLORE the Question. (11 mins. 36 secs. for DVD segment)

1. Show the *EVO* DVD, Question 10, Why should anyone care about evolution?

2. Provide time for students' questions and discussion. If necessary, use the *Look-Again* markers and return to the scientists' explanations. *Look-Again* markers will direct viewers to the following clips:
   - health, environmental degradation
   - cultural evolution
   - biotechnology dangers
   - conclusion, Dalai Lama quote

EXPLAIN the Reason to Care About Evolution. (40 mins.)

1. Tell the students that this is a final activity, and they are each going to prepare a short essay based on their respective answers to the question.

2. Indicate the following requirements for their essays:
   - Provide a clarifying statement of evolution. For example, students might go back to the first *EVO* question and formulate a statement such as the following: Biological evolution provides an explanation for the diverse forms of life and the commonalities among those diverse species. It also provides a scientific story of the history of life on Earth.
   - Identify one specific example of an application of evolution. The examples might include health, antibiotic resistance, or invasive species.
   - Include a quote from one of the scientists in the *EVO* DVD.
   - Cite two sources from a web search of the example.

3. Explain that they are to argue for their positions and support their arguments using the quotes and two sources.

ELABORATE the Reason to Care About Evolution.

1. The explanation developed by the students should include an elaboration. Specifically, the elaboration is the application of an understanding of evolution to a contemporary situation such as health or the environment.

EVALUATE Students' Learning.

Use the following criteria for evaluating students' essays and work:

Exceeds expectations: All outcomes are met, work goes beyond expectations and demonstrates exceptional understanding.

Meets expectations: Work indicates that student understands major issues and concepts.

Below expectations: Student work does not meet the criteria. It may be incomplete or may not address key issues or concepts of the learning outcomes.

Learning Outcomes	Evaluation
Synthesize knowledge about biological evolution	
Construct an argument using evidence	
Demonstrating communication and self-management skills	
Demonstrate systems thinking	
Overall evaluation for this lesson	

AUTHOR BIOGRAPHIES

Rodger W. Bybee (curriculum developer) was chair of the Programme for International Student Assessment (PISA) 2006 Science Expert Group. He is a former executive director of the Biological Sciences Curriculum Study (BSCS) and of the Center for Science, Mathematics, and Engineering Education at the National Academies of Science, in Washington, D.C. His major areas of work have included scientific literacy, scientific inquiry, the design and development of school science curricula, and the role of policy in science education. He received his PhD from New York University and his BA and MA degrees from the University of Northern Colorado. In 1989, he was recognized as one of the 100 outstanding alumni in the history of the University of Northern Colorado. In 1998, NSTA presented Dr. Bybee with the Distinguished Service to Science Education Award, and in 2001, he received the Robert H. Carleton Award, NSTA's highest honor, for national leadership in science education.

John Feldman (filmmaker) is a pioneering independent filmmaker whose work as a director of independent dramatic films and documentaries has earned him numerous awards, including the New American Cinema Award from the Seattle International Film Festival. His feature films *Alligator Eyes* (1990), *Dead Funny* (1995), and *Who the Hell Is Bobby Roos?* (2002) have enjoyed considerable international acclaim. He is highly regarded as a writer and director of short nonfiction films and has developed a reputation for his ability to explain complex ideas to a lay audience in a visually engaging way. He is currently focusing his efforts on educational films and revitalizing the use of film in the classroom. John has an MFA in filmmaking from Temple University and a BA in biology from the University of Chicago.

REFERENCES AND RESOURCES FOR USING *EVO*

REFERENCES

Alters, B. J., and S. M. Alters. 2001. *Defending evolution in the classroom*. Sudburg, MA: Jones and Bartlett Publishers.

Biological Sciences Curriculum Study (BSCS). 2009. *The biology teachers handbook, 4th edition*. Arlington, VA: NSTA Press.

Biological Sciences Curriculum Study (BSCS). 2005. *The nature of science and the study of biological evolution*. Arlington, VA: NSTA Press.

Bybee, R. W., ed. 2004. *Evolution in perspective*. Arlington, VA: NSTA Press.

Bybee, R. W., ed. 2006. *Evolutionary science and society: Activities for the classroom*. Colorado Springs, CO: BSCS.

Cracraft J., and R. W. Bybee, eds. 2005. *Evolutionary science and society: Educating a new generation*. Proceedings of the BSCS, AIBS Symposium, Nov. 2004, Chicago. Colorado Springs, CO: BSCS.

Darwin, C. (1859) 1964. *On the origin of species: A facsimile of the first edition with an introduction by Ernst Mayr*. Cambridge, MA: Harvard University Press.

Eldredge, N. 2005. *Darwin: Discovering the tree of life*. New York: W. W. Norton.

Goldschmidt, T. 1996. *Darwin's dreampond: Drama in Lake Victoria*. Cambridge, MA: MIT Press.

Mayr, E. 2000. Darwin's influence on modern thought. *Scientific American* 283 (1): 78–83.

National Academy of Sciences (NAS). 2008. *Science, evolution, and creationism*. Washington, DC: National Academies Press.

National Research Council (NRC). 1996. *National science education standards*. Washington, DC: National Academies Press.

National Research Council (NRC). 2011. *A Framework for K–12 science education: Practices, crosscutting concepts, and core ideas*. Washington, DC: National Academies Press.

Pigliucci, M. 2002. *Denying evolution: Creationism, scientism, and the nature of science*. Sunderland, MA: Sinauer Associates.

Roughgarden, J. 2006. *Evolution and Christian faith: Reflections of an evolutionary biologist*. Washington, DC: Island Press

Sagan, C. 1993. *Broca's brain: Reflections on the romance of science*. New York: Random House.

Scott, E. C. 2004. *Evolution vs. creationism: An introduction*. Westport, CT: Greenwood Press.

Skehan, J. W., and C. B. Nelson. 2000. *The creation controversy and the science classroom*. Arlington, VA: NSTA Press.

RESOURCES

American Association for the Advancement of Science (AAAS). 1989. *Science for all Americans*. Washington, DC: AAAS.

American Association for the Advancement of Science (AAAS) Project 2061. 2006. *Evolution on the front line: An abbreviated guide for teaching evolution from project 2061*. Washington, DC: AAAS.

American Association for the Advancement of Science (AAAS). C. Baker and J. Miller., eds. 2006. *The evolution dialogues: Science, Christianity and the quest for understanding.* Washington, DC: AAAS.

American Museum of Natural History (AMNH). 2009. *Evolution: Resource DVD.* New York: AMNH.

Ayala, F. J. 2007. *Darwin's gift: To science and religion.* Washington, DC: John Henry Press.

Baker, C., and J. Miller, eds. 2006. *The evolution dialogues: Science, Christianity, and the quest for understanding.* Washington, DC: AAAS.

Barbour, I. G. 2000. *When science meets religion: Enemies, strangers, or partners?* New York: HarperCollins.

Bishop, B. A. and C. W. Anderson. 1990. Student conceptions of natural selection and its role in evolution. *Journal of Research in Science Teaching* 27 (5): 415–427.

Biological Sciences Curriculum Study (BSCS). 2007. *Evolution and the environment: Activities for the classroom.* Colorado Springs, CO: BSCS.

Biological Sciences Curriculum Study (BSCS). 2004. *Galápagos: An inquiry into biological evolution.* Arlington, VA: NSTA Press.

Cherif, D., G. Adams, and J. Loehr. 2001. What on earth is evolution? The geological perspective of teaching evolutionary biology effectively. *The American Biology Teacher* 63 (8): 569–589.

Jensen, J. E. 2008. *NSTA tool kit for teaching evolution.* Arlington, VA: NSTA Press.

Larson, E. J. 2004. *Evolution: The remarkable history of a scientific theory.* New York: Random House.

Larson, E. J. 2001. *Evolution workshop: God and science on the Galápagos Islands.* New York: Basic Books.

Margulis, L., and D. Sagan. 2002. *Acquiring genomes: A theory of the origins of species.* New York: Basic Books.

Mayr, E. 1993. *One long argument: Charles Darwin and the genesis of modern evolutionary thought.* Cambridge, MA: Harvard University Press.

Mayr, E. 2001. *What evolution is.* New York: Basic Books.

Miller, K. R. 2000. *Finding Darwin's god: A scientist's search for common ground between God and religion.* New York: HarperCollins.

Moore, J. A. 2002. *From genesis to genetics: The case of evolution and creationism.* Berkeley, CA: University of California Press.

Quammen, D. 2006. *The reluctant Mr. Darwin: An intimate portrait of Charles Darwin and the making of his theory of evolution.* New York: W. W. Norton.

The New York Botanical Garden. 2008. *Darwin's garden: An evolutionary adventure.* Bronx, NY: The New York Botanical Garden.

Understanding Evolution at *http://evolution.berkeley.edu*

Understanding Science at *http://www.understandingscience.org*

Weiner, J. 1994. *The beak of the finch: The story of evolution in our time.* New York: Vintage Books.

APPENDIX 1.1

From *Zoological Philosophy*, Jean Lamarck

The environment affects the shape and organization of animals, that is to say that when the environment becomes very different, it produces in course of time corresponding modifications in the shape and organization of animals.

If a new environment, which has become permanent for some race of animals, induces new habits in these animals, that is to say, leads them into new activities which become habitual, the result will be the use of some one part in preference to some other part, and in some cases the total disuse of some part no longer necessary.

Nothing of all this can be considered as hypothesis or private opinion; on the contrary, they are truths which, in order to be made clear, only require attention and the observation of facts. …

The frequent use of any organ, when confirmed by habit, increases the functions of that organ, leads to its development, and endows it with a size and power that it does not possess in animals which exercise it less.

We have seen that the disuse of any organ modifies, reduces, and finally extinguishes it. I shall now prove that the constant use of any organ, accompanied by efforts to get the most out of it, strengthens and enlarges that organ, or creates new ones to carry on the functions that have become necessary.

The bird which is drawn to the water by its need of finding there the prey on which it lives, separates the digits of its feet in trying to strike the water and move about on the surface. The skin which unites these digits at their base acquires the habit of being stretched by these continually repeated separations of the digits; thus in course of time there are formed large webs which unite the digits of ducks, geese, etc. as we actually find them.

It is interesting to observe the result of habit in the peculiar shape and size of the giraffe; this animal, the largest of the mammals, is known to live in the interior of Africa in places where the soil is nearly always arid and barren, so that it is obliged to browse on the leaves of trees and to make constant efforts to reach them. From this habit long maintained in all its race, it has resulted that the animal's fore-legs have become longer than its hind legs, and that its neck is lengthened to such a degree that the giraffe, without standing up on its hind legs, attains a height of six metres (Nearly twenty feet).

Lamarck, J. *Philosophie zoologique*. Translated by H. Elliott. London: Macmillan, 1914.

APPENDIX 1.2

From *On the Tendency of Varieties to Depart Indefinitely from the Original Type,* Alfred Russel Wallace

THE STRUGGLE FOR EXISTENCE

The life of wild animals is a struggle for existence. The full exertion of all their faculties and all their energies is required to preserve their own existence and provide for that of their infant off-spring. The possibility of procuring food during the least favorable seasons and of escaping the attacks of their most dangerous enemies are the primary conditions which determine the existence both of individuals and of entire species.

THE LAW OF POPULATION OF SPECIES

The numbers that die annually must be immense; and as the individual existence of each animal depends upon itself, those that die must be the weakest—the very young, the aged, and the diseased—while those that prolong their existence can only be the most perfect in health and vigor, those who are best able to obtain food regularly and avoid their numerous enemies. It is "a struggle for existence," in which the weakest and least perfectly organized must always succumb.

USEFUL VARIATIONS WILL TEND TO INCREASE, UNUSEFUL OR HURTFUL VARIATIONS TO DIMINISH

Most or perhaps all the variations from the typical form of a species must have some definite effect, however slight, on the habits or capacities of the individuals. Even a change of color might, by rendering them more or less distinguishable, affect their safety; a greater or less development of hair might modify their habits. More important changes, such as an increase in the power or dimensions of the limbs or any of the external organs, would more or less affect their mode of procuring food or the range of country which they could inhabit. It is also evident that most changes would affect, either favorable or adversely, the powers of prolonging existence. An antelope with shorter or weaker legs must necessarily suffer more from the attacks of the feline carnivore; the passenger pigeon with less powerful wings would sooner or later be affected in its powers of procuring a regular supply of food; and in both cases the result must necessarily be a diminution of the population of the modified species.

Wallace, A. R., On the tendency of varieties to depart indefinitely from the original type. *Journal of the Proceedings of the Linnean Society.* August 1858.

APPENDIX 1.3

From *Introduction, On the Origin of Species*, Charles Darwin

When on board HMS *Beagle*, as naturalist, I was much struck with certain facts in the distribution of the inhabitants of South America, and in the geological relations of the present to the past inhabitants of that continent. These facts seemed to me to throw some light on the origin of species—that mystery of mysteries, as it has been called by one of our greatest philosophers. On my return home, it occurred to me, in 1837, that something might perhaps be made out on this question by patiently accumulating and reflecting on all sorts of facts which could possibly have any bearing on it.

My work is now nearly finished; but as it will take me two or three more years to complete it, and as my health is far from strong, I have been urged to publish this Abstract. I have more especially been induced to do this, as Mr. Wallace, who is now studying the natural history of the Malay archipelago, has arrived at almost exactly the same general conclusions that I have on the origin of species. Last year he sent to me a memoir on this subject, with a request that I would forward it to Sir Charles Lyell, who sent it to the Linnean Society, and it is published in the third volume of the Journal of that Society. Sir C. Lyell and Dr. Hooker, who both knew of my work—the latter having read my sketch of 1844—honoured me by thinking it advisable to publish, with Mr. Wallace's excellent memoir, some brief extracts from my manuscript.

In considering the Origin of Species, it is quite conceivable that a naturalist, reflecting on the mutual affinities of organic beings, on their embryological relations, their geographical distribution, geological succession and other such facts, might come to the conclusion that each species had not been independently created, but had descended, like varieties, from other species. Nevertheless, such a conclusion, even if well founded, would be unsatisfactory, until it could be shown how the innumerable species inhabiting this world have been modified, so as to acquire that perfection of structure and coadaptation which most justly excites our admiration. Naturalists continually refer to external conditions, such as climate, food, etc., as the only possible cause of variation. In one very limited sense, as we shall hereafter see, this may be true; but it is preposterous to attribute to mere external conditions, the structure, for instance, of the woodpecker, with its feet, tail, beak, and tongue, so admirably adapted to catch insects under the bark of trees. In the case of the misseltoe, which draws its nourishment from certain trees, which has seeds that must be transported by certain birds, and which has flowers with separate sexes absolutely requiring the agency of certain insects to bring pollen from one flower to the other, it is equally preposterous to account for the structure of this parasite, with its relations to several distinct organic beings, by the effect of external conditions, or of habit, or of the volition of the plant itself. …

It is, therefore, of the highest importance to gain a clear insight into the means of modification and coadaptation. At the commencement of my observations it seemed to me probable that a careful study of domesticated animals and of cultivated plants would offer the best chance of making out this obscure problem. Nor have I been disappointed; in this and in all other perplexing cases I have invariably found that our knowledge, imperfect though it be, of variation under domestication, afforded the best and safest clue. I may venture to express my conviction of the high value of such studies, although they have been very commonly neglected by naturalists.

No one ought to feel surprise at much remaining as yet unexplained in regard to the origin of species and varieties, if he makes due allowance for our profound ignorance in regard to the mutual relations of all the beings which live around us. Who can explain why one species ranges widely and is very numerous, and why another allied species has a narrow range and is rare? Yet these relations are of the highest importance, for they determine the present welfare, and, as I believe, the future success and modification of every inhabitant of this world. Still less do we know of the mutual relations of the innumerable inhabitants of the world during the many past geological epochs in its history. Although much remains obscure, and will long remain obscure, I can entertain no doubt, after the most deliberate study and dispassionate judgment of which I am capable, that the view which most naturalists entertain, and which I formerly entertained—namely, that each species has been independently created—is erroneous. I am fully convinced that species are not immutable; but that those belonging to what are called the same genera are lineal descendants of some other and generally extinct species, in the same manner as the acknowledged varieties of any one species are the descendants of that species. Furthermore, I am convinced that Natural Selection has been the main but not exclusive means of modification.

Darwin, C. *On the origin of species by means of natural selection.* London, 1859.

INDEX

NATIONAL SCIENCE TEACHERS ASSOCIATION